Fight, Flight, Mimic

Fight, Flight, Mimic

Identity Mimicry in Conflict

Edited by

Diego Gambetta
Thomas Hegghammer

OXFORD
UNIVERSITY PRESS

OXFORD
UNIVERSITY PRESS

Great Clarendon Street, Oxford, OX2 6DP,
United Kingdom

Oxford University Press is a department of the University of Oxford.
It furthers the University's objective of excellence in research, scholarship,
and education by publishing worldwide. Oxford is a registered trade mark of
Oxford University Press in the UK and in certain other countries

Published in the United States of America by Oxford University Press
198 Madison Avenue, New York, NY 10016, United States of America

British Library Cataloguing in Publication Data
Data available

Library of Congress Control Number: 2024932017

ISBN 9780198739470

DOI: 10.1093/9780191802454.001.0001

Printed and bound by
CPI Group (UK) Ltd, Croydon, CR0 4YY

MIX
Paper | Supporting
responsible forestry
FSC® C013604

Contents

List of Contributors   vi

1. **The Theory and History of Mimicry in Conflict**   1

   Diego Gambetta and Thomas Hegghammer

2. **Strategic Dynamics of Social Mimicry**   27

   James D. Fearon

3. **Can You Trust Anyone on Jihadi Internet Forums?**   50

   Thomas Hegghammer

4. **The Code Word Conundrum in the Northern Ireland Conflict**   90

   Heather Hamill

5. **The Red Brigade Signature: Mimic-proofing Claims of Political Violent Actions, Italy, 1969–1980**   114

   Valeria Pizzini-Gambetta and Diego Gambetta

6. **Where Are the Mimics When Passing Seems Easy?: The Rwandan Genocide in Comparative Perspective**   152

   David D. Laitin

7. **'Trademark Wars': Naxals versus Criminal Extortionists in India**   176

   Pavan Mamidi

8. **Mimicry and Its Double in the Iraqi Civil War**   205

   Stephen Holmes

Index   247

List of Contributors

James D. Fearon PhD is Professor of Political Science, Stanford University, Stanford, CA, US. https://orcid.org/None

Diego Gambetta PhD, FBA is Chair of Social Science, Collegio Carlo Alberto, Turin, Italy. https://orcid.org/0000-0002-1655-6615

Heather Hamill DPhil is Professor of Sociology, Department of Sociology, University of Oxford, UK. https://orcid.org/0000-0002-0301-7057

Thomas Hegghammer PhD is Senior Research Fellow, All Souls College, Oxford University, Oxford, UK. https://orcid.org/0000-0001-6253-1518

Stephen Holmes PhD is Professor of Law, New York University School of Law, New York, NY, US. https://orcid.org/0009-0002-7768-3818

David D. Laitin PhD is Professor of Political Science, Stanford University, CA, US. https://orcid.org/0000-0002-7911-7089

Pavan Mamidi DPhil is Director, CSBC, Ashoka University, India. https://orcid.org/None

Valeria Pizzini-Gambetta PhD is Independent Researcher, n\a, Turin, Italy. https://orcid.org/0009-0008-6775-9376

1

The Theory and History of Mimicry in Conflict

Diego Gambetta and Thomas Hegghammer

Midas, pretending that he was going to perform a solemn sacrifice to the great gods, led out the Phrygians by night as if in a procession, with flutes, and timbrels, and cymbals; but each of them at the same time secretly carried swords. The citizens all left their houses to watch the procession; but the musical performers drew their swords, slew the spectators as they came out into the streets, took possession of their houses, and set up Midas as their ruler.

Polynaeus, *Stratagems* 7:5 (from Shepherd 1793: 264), about the 7th century BC King Midas of Phrygia

Warren Street bomber Yassin Omar was defined by one image throughout his six-month trial—that of him fleeing London after July 21 dressed as a woman wearing a burka. Shadowy CCTV footage the following day showed a lanky 1.83-metre figure in a long black robe passing through Golders Green coach station and stepping off at Dig-beth in Birmingham several hours later. To add to the bizarre scene, the 26-year-old—who shaved his arms to look more like a woman—could be clearly seen holding a white handbag.

***Guardian*, 9 July 2007, about Yassin Omar, who evaded police for six days after an attempted suicide bombing in London on 21 July 2005**

Of all the dirty tricks of war, one of the most potent and pernicious is to lie about who you are.[1] Posing as someone your enemy trusts or is indifferent to, can give you access to his inner quarters, allowing you to steal his secrets, or hit him where it really hurts. Conversely, a good disguise can get you out of

[1] We are grateful to Juan Masullo for helping to edit the entire book manuscript and providing numerous suggestions for improvement. We also thank Noel Malcolm, Wolfgang Ernst, and Målfrid Braut-Hegghammer for sharing sources and ideas.

Diego Gambetta and Thomas Hegghammer, *The Theory and History of Mimicry in Conflict*. In: *Fight, Flight, Mimic*. Edited by: Diego Gambetta and Thomas Hegghammer, Oxford University Press. © Diego Gambetta and Thomas Hegghammer (2024). DOI: 10.1093/9780191802454.003.0001

a dangerous situation and avoid the harm that the enemy would otherwise inflict on you. Deceptive identity mimicry—or mimicry for short—is a tool of war as potent as any weapon or fortification.

If mimicry has not yet received much attention as a feature of conflict, it is partly because it does not say 'boom'. If it succeeds, it may remain undetected, and if warring parties deter it by taking precautions against it, it remains unrealized, a mere counterfactual. And when it does manifest itself, it takes a myriad different forms, making it an unwieldy topic of study. No stable and definite rules dictate what makes or breaks a mimicry attempt, for it is always a dynamic game between at least two parties, and context-specific details matter greatly. Yet, as we shall see in this book, mimicry shapes the behaviour and organization of conflict actors to an extent that has hitherto not been appreciated.

Identity mimicry occurs in many contexts as a deeply subversive ingredient in human interaction. We know it from as ancient a text as Genesis 27 from the Old Testament, in which Jacob pretends to be his hairy brother Esau by wearing goatskin on his neck and arms, thereby tricking his blind father Isaac into blessing him. We encounter it in our daily lives through email scams, phone calls from fake representatives of Microsoft Support, and Tinder dates who were nothing like what their profiles promised. Banks and businesses experience it through relentless fraud attempts involving fake identities, and diplomats and politicians face it in espionage threats. Most of us are cautious with strangers because we realize that trusting an impostor can have dire consequences. Perhaps that is why children are warned of it in *Grimms' Fairy Tales*, where Little Red Riding Hood gets eaten by a wolf dressed as her grandmother.

It is not only bad guys (and wolves) who mimic, for morality is orthogonal to mimicry. Mimicry often serves valiant causes, whether offensively—think infiltration of the mob, as in the case of FBI special agent Joseph Pistone, aka Donnie Brasco (see the film by the same name)—or defensively, as with Jews passing off as Christian Poles to escape the Gestapo during the Second World War. Deceptive identity signalling is simply a tactic, but can be a very powerful one at that.

Rich societies have developed elaborate systems and institutions to prevent and mitigate fraudulent identity signalling. Passports, passwords, shibboleths, register offices, biometric databases, intelligence services, and other systems permeate modern society to the point where we consider them as facts of life. These days it is perhaps only when we lose our wallets or smartphones—and become economically paralyzed for lack of identification tools—that we experience the real impact of these systems. But their effectiveness makes it similarly easy to underestimate the threat posed by mimicry to the social and economic order. In wealthy and stable countries, people experience identity fraud relatively infrequently, because verification systems are preventing and deterring it.

Not all communities enjoy these stabilizing mechanisms. In the criminal underworld, for example, you cannot go to the police when you have been conned, and checking a potential partner's criminal credentials is quite laborious. In such contexts, we can often better see the destructive potential of mimicry and the complex trust games it engenders. This was the motivation for *Codes of the Underworld*, Gambetta's (2009a) exploration of interpersonal trust in the criminal world, as well as for several other studies of domains where trust is precarious (Donath 1998; Gambetta and Hamill 2005).

In this book, we look at mimicry in an even more delicate type of situation, namely, civil wars. Here, official verification systems are often lacking, just as in the criminal world, but the stakes are even higher. What happens to mimicry games when we move from crime to politics and when we ramp up the stakes to being about life and death? Is tricking someone for cash fundamentally the same as deceiving them to kill or survive? Will high-stakes settings generally see more mimicry (because mimics have more to gain) or will they see less (because targets of mimicry will be more determined not to be duped)? We do not really know the answer to these questions, because mimicry in conflict has not yet received in-depth scholarly attention. This is not for lack of data, for there is ample anecdotal evidence of wartime mimicry in the sources, as we shall see. However, it takes in-depth case knowledge to spot and properly understand mimicry games, which is why we recruited specialist contributors and made this an edited volume.

The aim of this book is to make at least three contributions: first, to deepen our knowledge of mimicry by studying it in a new context; second, to refine our understanding of deception in war by highlighting mimicry as a peculiar subcategory of deception; and third, to improve our understanding of civil wars by drawing attention to an underappreciated dynamic within them. This opening chapter will clarify concepts, lay out a theoretical framework, and provide historical perspective before handing it over to the contributors and their case studies. But first: what exactly are we talking about?

How Identity Mimicry Relates to Other Means of Deception

Mimicry is a form of deception.[2] Drawing loosely on the Merriam-Webster Dictionary definition, deception is the act of making someone believe that a state of affairs is true when it is not. It is useful to think about deception as having both a strategic and a tactical dimension. The strategic dimension pertains to the nature of the state of affairs being misrepresented. This can vary: sometimes we deceive about basic facts about the world, such as when a cheating husband tries to keep his affair hidden from his wife. At other times we misrepresent our intentions, for example, when an Ebay vendor puts something up for sale with the intention of cashing the advance payment without ever delivering the product.

The tactical dimension of deception relates to the means by which the misrepresentation is achieved. A basic method is plain lying, where we use words to misrepresent the world. Another ideal type tactic is false pretense, which involves non-verbal means to distort other people's perception of the world. There are many other conceivable means of deception: forgery, imitation, disguise, concealment, calculated silence, and more. These tools are often combined for a strategic deceptive purpose: for example, the cheating husband may lie, conceal, use false pretense, or even forgery (of, say, a hotel bill) to keep his affair hidden from his partner.

It is when the state of affairs being manipulated is the identity of the person or her objective belonging to a particular category or group that we can speak of mimicry. Here we mean identity in the sense of human individuality—i.e. a person's name and other facts about who he or she is—not identity in the sense of self-identification. A closely related term is impersonation, but mimicry includes more than pretending to be a specific other individual. In a typical mimicry episode, the mimic pretends to be a different person or a member of a group to which he does not belong to obtain some advantage he would not otherwise obtain. Deceptive mimicry, therefore, is a particular form of deception that involves the strategic manipulation of identity. It is usually a composite phenomenon built on more basic tactics of deception. One can lie about one's identity, forge a document or business card, disguise oneself with clothes or make-up, or imitate the voice, gait, or other feature of the person or group being mimicked.

These concepts can be meaningfully transposed to the domain of war and help us distinguish between different ideal types of wartime deception. There

---

[2] For more on deception in general, see for example: Bell 1982; Handel 1982; Bell and Whaley 1991; Caddell 2004; Martin 2009; and Clark and Mitchell 2018.

is straightforward lying, for example when a party promises another party safe passage but breaks the promise and attacks when the other is vulnerable (also known as perfidy). There is false pretense, for example, when a general dispatches a force to location A to make the enemy think he will attack there, when the main attack really involves another force in location B. Deceptive mimicry, by contrast, is when, for example, you send an infiltrator into enemy ranks or dress up your troops in civilian clothes to avoid detection.

Linking mimicry tightly to identity also helps us draw less obvious distinctions. For example, since identity in this context is a primarily human characteristic, it is not mimicry when a person pretends to look like or hide behind an innate object.[3] The soldiers inside the Trojan horse were not engaged in regular mimicry, but in a form of false pretense, stealth, or camouflage.[4] They were veiling their real identities rather than assuming other people's identities. Similarly, making one type of animal look like another (for example, dressing donkeys as cavalry horses to make your army seem bigger) or setting up cardboard imitations of tanks and planes are not mimicry, although one could argue it is conceptually close.[5] Similarly, espionage often involves identity manipulation, but not always. Spying can sometimes be about covertly collecting information through observation and eavesdropping, or about managing informants, in which case the intelligence operative does not necessarily need to pass him- or herself off as someone else.

Sometimes mimicry is central to what the mimic is trying to achieve, such as when a Hamas suicide bomber dresses up as an Orthodox Jew to detonate his belt inside a synagogue. At other times, the mimicry is but a cog in a larger wheel. In the famous Operation Mincemeat in 1943, British intelligence took the dead body of a Welsh homeless man named Glyndwr Michael and dressed it up as the fictitious Royal Marine Captain William Martin and dropped it off outside the southern coast of Spain. To the body they attached fake documents suggesting the Allies were planning to invade Greece and Sardinia—a ploy to draw attention away from the real plan, namely, the invasion of Sicily. Here, mimicry was an element in a larger false pretense operation.[6]

---

[3] Animals can have individual identities, but it is usually because humans give it to them. In biology, mimicry refers to imitation across species, not within them. To our knowledge, intra-species mimicry is not a thing in the animal world. But the converse—humans mimicking animals in the classical biological sense—can occur; for example, in the form of a person donning a gorilla suit or whistling bird-like sounds.

[4] Conceptually, the Trojan horse ploy is close to the notion of crypsis in biology; the ability of an animal or plant to avoid being seen by other animals.

[5] The old tactic of deception through props is alive and well: see e.g. Jan Lopatka, 'Czech company sees boom in market for fake tanks, HIMARS' *Reuters*, 6 March 2023.

[6] Although Operation Mincemeat is famous—not least after its depiction in a film in 2021—it is actually a rather unusual mimicry episode insofar as it is instigated by the model and involves a mimic who is dead.

Identifying the constituent parts of deception helps us see that some phenomena in the world of deception and disinformation are not distinct types of deception but aggregations of smaller deception episodes. For example, a disinformation campaign arguably rests on an aggregation of mimicry episodes, because it relies on multiple fake news producers or Russian troll factories passing themselves off as reputable news organizations. This means that the study of deceptive mimicry, even though it often involves delving into the nitty-gritty of individual interactions, may hold the key to understanding larger phenomena such as information warfare, disinformation campaigns, and the like.

Mimicry is thus a subcategory of deception, but it is one of the most important ones at that. In some sense it is the master deception, as it is upstream to several other deceptive tactics. Once you have successfully passed under another identity, it becomes much easier to deploy other forms of trickery and subterfuge. Moreover, mimicry follows a distinct logic that sets it apart from other types of deception.

How Identity Signalling Works and How Mimics Can Exploit It

The theoretical study of mimicry has produced several concepts that help us understand its mechanics and relationship to identity signalling. Mimicry is a term derived from biology (Carpenter and Ford 1933; Wickler 1968; Forbes 2009; Sun 2023), and in the deceptive variety that interests us here, consists of passing off either as a different individual or as a member of a group to which one does not belong. Each mimicry episode has three protagonists, often nimbly defined as mimic, model, and dupe. In order to improve either their aggressive or defensive payoffs the mimics aim to make the dupes believe that they are the model (Gambetta 2005).

Here it is important to set identity mimicry in the broader context of deceptive mimicry generally. Humans can feign certain dispositions—curiosity, excitement—or physical and emotional states—headaches, grief, or orgasm—by imitating the looks, postures, words, or level and type of perceivable activity directly associated with these states. They can deceitfully pretend to possess an unobservable but desirable property k, such as benevolence or honesty, by adopting a sign m that is believed to be associated with k, e.g. looking people in the eyes to persuade them of one's honesty. The model for the mimic in these cases is not a specific agent or group of agents, but a generic state or property and its purported manifestations. This is a form of false pretense and is not the type of deceptive mimicry that concerns us here.

A great deal of human signalling takes place indirectly by signalling one's identity, both as a specific individual, or as a member of a group or category. After we encounter an individual or group member and experience dealing with them, we form and retain an idea of whether this person or group has or lacks the k property that interests us—it creates a 'reputation', whether positive or negative. Identity signalling enables the signaller to exploit a positive reputation. In identity signalling, instead of using a two-layered inferential structure ($m \rightarrow k$), we use a three-layered structure, $g \rightarrow i \rightarrow k$, where i denotes identity and g denotes a sign of identity, or signature. If persons or group members are reidentifiable by some signature, the next time we meet them we infer the presence or absence of k via their identity i.

The reidentification of a signature, however, can be problematic. This is because the fact that someone has a certain reputation is an unobservable property of that person. For example, Armani has a reputation for selling well-designed clothes, but to exploit this reputation, the seller must convince customers that she is Armani. Islamic State has a reputation for carrying out its threats against hostages, but to exploit this reputation a group of kidnappers must convince governments that they really belong to Islamic State. When a model signals qualities via her or his identity, the threat of mimicry of k through m is replaced by the threat of mimicry of i through g. Much of human deceptive mimicry exploits signalling via identity, and our interest is mostly in this case.

Note also that, while the mimic and the dupe are always ultimately one or more individuals (one can only intentionally fool or be fooled as an individual), the model may be either an individual or the features of the members of a group or category (or, in the case of camouflage, even the features of an environment).[7] In short, while the mimic and the dupe require a mind, the model does not necessarily require one. Depending on the mimicry 'system', the model may be inert or indifferent, but may also either fight against or collude with the mimic. In any case, the dupe, as we shall see, is not always so easily duped.

---

[7] Camouflage can be conceptualized as a case of *negative mimicry*. There are often signs that are likely to be interpreted by the signal receiver, rightly or wrongly, as indicating the absence of a property k of the signaller. Both an honest signaller with k who expects to be unjustly perceived if he displays such a sign m, and an opportunist non-k who is afraid of being detected if he does, have a reason to camouflage. That is, they take steps *not* to show m. Camouflaging can be considered as a special case of mimicking, since the strategy of camouflaging non-k-ness by suppressing m is just that of mimicking k through displaying the notional sign '*no m*'(Gambetta 2005).

Identity Mimicry as a Game

Mimicry is best understood theoretically as a strategic game because the outcome for each player depends on the actions of the other players. James Fearon's contribution to this volume (Chapter 2) illustrates this point forcefully. At the same time, mimicry is a communication act that belongs to the class of signalling games—which differs from other classical games such as the Prisoner's Dilemma—and is therefore worth examining in some depth.

In a one-shot case, the mimic moves by doing his mimicry act, and either the dupe falls for it and does what the mimic wants the dupe to do (and that the dupe would not do if they knew they were facing a mimic); or does not fall for it and simply does not do that, and the game ends there. Imagine someone looking like a veteran displaying a disability and begging on the street: some people will decide he is pretending and move on, others will believe he is genuine and consider giving him money.

The decision process of the potential dupe has two main stages. First of all she decides whether the base-rate probability of deception in her environment is high enough to consider cheating an actual risk. In the affirmative case, she then evaluates whether the signals that identify a person as X or as a member of group Y are genuine by filtering them through the principle of signalling theory: 'considering what they would gain from deceiving me, is the cost affordable only to someone who is truly x or y, while is hardly affordable to a mimic'? Only if those signals are too costly to be mimicked given the plausible gains, the potential dupe believes that they are facing an honest signaller (Gambetta 2005; Gambetta and Hamill 2005).

Imagine someone rings the bell and, when asked via the intercom 'who's there?', replies 'postman!'. When do we believe that to be the truth? Uttering 'postman' in an intercom that does not have a video camera and poorly transmits the voice is cheap, almost anyone could do that; so the answer further depends on what the dupe expects that a mimic could gain from passing off as the postman. If dangerous intruders are believed to be frequent, then the dupe will not take the 'postman' claim as true. The game can end there and simply the door will not be opened, or it can go on: the unconvinced dupe can probe the 'postman' claim further. Contrary to what their moniker in biology suggests, a dupe is rarely a passive victim of mimicry, but rather an intelligent receiver of signals aware of the threat of deception. In conflict situations, when stakes are highest and anyone could be a threat, the dupes are likely to keep their guard high. This includes deploying active countermeasures and tactics to weed out mimics. The development of such tactics involves no less creativity and cunning than the mimicry efforts they are designed to stop. Human collectives often collate screening tactics into entire systems and build institutions designed to spot mimicry and ultimately deter it. If the dupe suspects to be facing a fraudster, they could probe the identity by asking further questions based on local knowledge, or if the threat is recurrent then some further screening device is likely to be in place—a code number instead of just a bell, a video camera to inspect the facial features of the 'postman', or traditionally an attentive concierge that controls a second internal door.

But then mimics may respond and invest more resources to counter the filtering tactics and try outfoxing the systems. The mimic can don a postman's uniform for instance, carry a parcel and such like. Mimicry then becomes a race, a crucial mode in which the predator–prey (mimic–dupe) dynamic is realized. One can understand from this perspective that mimicry can act as a powerful engine of change: cognitive, technological, and institutional.[8]

Facial features are a good example upon which to reason further. They are near-impossible to reproduce in natural circumstances, harder than the voice, the gait, the age, or the handwriting of another person. The cost of reproducing someone's face in three dimensions and in human flesh is extremely high, hence displaying one's face is a robust signal of one's identity, at least within the same ethnic group. If the dupe can somehow see the face of the postman whom she knows from previous interaction, she will know whether he is telling the truth.

Humans are generally good at discerning and remembering faces. But our ability to identify a given person through their face rests on a host of other conditions, some of which can be exploited or manipulated by a mimic. For instance, fraudsters often target elderly people, for they are more likely to have weak eyesight and patchy memory. Mimics may also wear hats, scarves, or glasses to shade their features. But if we are in possession of good cognitive skills and see the face of person X clearly enough, we can be near-certain that that face truly belongs to X, in which X is the name of a person and all the features we associate with her. The face of an individual displays such a unique constellation of features that it was only relatively late in the digital era that we got computer vision models capable of reliable facial recognition.

The power of the face as an identifier now only applies in an analogue, three-dimensional environment, for in the digital world, the cost of

---

[8] The fundamental contribution of cheating to evolution generally is made by Sun (2023).

10

mimicking a face has been drastically reduced. We now have the technical tools to lift the face of a person off from a video recording and edit it in various ways to make it seem as if that person is saying things she never said. Artificial intelligence does the editing job for us and produces deep fakes of stunning accuracy.

All the above illustrates that mimicry opportunities are fluid, they can change as a result of changes in technology. As soon as our individual identifiers become detachable from our physical features and become man-made, even the face, this most robust of signals, is threatened as an identifier. But then again, people learn, and the race moves to a higher level. We can expect new forms of probing; maybe AI tools will be developed to screen out deep fakes and allow us to regain some provisional confidence that the screen is showing us the real face of X or the speaker broadcasting the true voice of Y. For the moment, we can share former US president Barack Obama's concern about 'the increasing sophistication of artificial intelligence and its ability to mimic public figures. He said only his wife was able to discern if a deep fake audio recording purporting to be him was actually him'.[9]

The most interesting dynamic of mimicry is endogenous to the relationship between the protagonists. In fact, the dynamics between mimics and dupes may have started a very long time ago. In evolutionary terms, it is reasonable to think that unique facial differences and the ability to recognize them have co-evolved within groups, and that what we now see as superficial natural variation is the result of the phenotypic hard labour of natural selection. Recognizing each other made it possible to build a reputation, and that in turn sustained our cooperation, which favoured our survival—we could more easily track down and punish free-riders, cheats, and others who hamper cooperation. Now, with the rise of AI deep fakes, we must find another solution to this problem in much less time.

So far we have only talked about mimics and dupes, but the model can also play a significant part in the mimicry game. Models can be indifferent and abstain from responding to a mimic that exploits their identifiers—this occurs if models have no interests, such as when they are fictional. In other mimicry systems, models collude with mimics as for instance when dictators hire doubles for self-protection. In yet other cases, the dupe may allow or even promote mimicry of their own group membership—in this case, the mimic is a shield for the model—as in the Second World War, when Polish members of the resistance helped Jews passing off as Poles to escape Nazi persecution (Finkel 2017; Braun 2019).

[9] Elias Visionary, 'Rupert Murdoch has fuelled polarization of society, Barack Obama says', the *Guardian*, 28 March 2023.

More often, however, models suffer detrimental effects as a result of mimics' actions. When the Norwegian politician Mímir Kristjánsson tweeted in 2022 that 'life consists of eight hours of work, eight hours of sleep, and eight hours of two-factor authentication', he was expressing the frustration many of us feel about modern digital life.[10] The rise of cumbersome authentication tools and procedures is a collateral damage imposed either by the self-defence tactics of the model that does not want to be mimicked or by those of the dupe. Models must sometimes also work harder to protect their reputation soiled by criminal mimics (a few chapters in this volume illustrate these cases). They are incentivized to erect barriers against mimicry and become de facto allies of dupes. Taxi drivers, for instance, form named cooperatives, share logos, coordinate the colours of their cars, display photos and phone numbers in the cab, in order to reassure passengers that they are hiring a bona fide driver and not a rapist posing as a taxi driver (Gambetta and Hamill 2005).

We can draw some interesting methodological lessons from studying mimicry. One is that seemingly small things can suddenly acquire a salient value. Critical signs can be relatively trivial, for example, the pronunciation of a single letter of a single word in the case of the Shibboleth or knowledge of cricket match results, the knowledge of which was used by British interrogators during the Second World War to uncover whether a suspect was a German spy (Gambetta 2009b). While signalling theory is a sharp tool, it often requires large amounts of intimately local knowledge to be applied efficiently.

Moreover, signalling knowledge is often short-lived, for two main reasons. One is that technological and other social transformations corrupt certain signals by making them too cheap to fool anyone. Another, more important reason is that people continuously learn how to mimic and how to protect themselves from it. For example, people are now more suspicious of unsolicited emails from stranded princes promising them millions in exchange for their bank details than they were in the early days of the Internet. The fickleness and highly diverse nature of signalling knowledge may be why the study of human mimicry has gone underdeveloped. Key signals are trivialized by being reported for their anecdotal value, and their disparate array has made the effort to unite them under the same analytical umbrella look impervious. The essential variable to look for is not the object that stands for a signal as such, but the differential cost of producing or displaying that

---

[10] Mímir Kristjánsson (@mimirk), *Twitter*, 26 September 2022 (https://twitter.com/mimirk/status/1,574,394,249,193,324,544).

object that different categories of people have. For example, a missing limb strengthens someone's claim to be a war veteran, not because of what limbs, present or absent, communicate to us as such, but because amputation would be a prohibitively high cost to bear for a healthy mimic.

Lastly, identity mimicry belongs to an interesting class of social phenomena whose latent threat affects society deeply without always being observable. Other examples include nuclear war, terrorism, plane crashes, and old diseases like the plague. In equilibrium these phenomena are so rare they seem not to exist. Yet, their near-invisibility is a result of the vast efforts that have gone into suppressing them. By the same token, mimicry can have a very large counterfactual importance. It can just be imagined and feared, or experienced in the past, and have led to countermeasures that are successful for long enough that their raison d'être is no longer apparent. For example, the airport security checks introduced after 9/11 seem tedious and unnecessary now that hardly anyone gets caught trying to hijack airplanes. But an important reason nobody gets caught is likely that the security checks are dissuading would-be terrorists from even trying. Deterrence can lead to underestimating the importance of mimicry, or even ignoring its relevance as a cause of institutional practices and technology deployment, and more generally of social evolution. In fact, the larger the importance the higher the chance that countermeasures have been put in place that made it (temporarily at least) disappear.

The struggle between mimics and countermimics is a constant in history. Eliav-Feldon (2012: 1) rightly inserts her fascinating book on identity fraud in seventeenth-century Europe into a 'long history of a contest between the forgers of identities and the creators of new and more efficient methods of identification, methods which in their turn bred new imaginative ways for evading the removal of masks'. Let us now take a closer look at that history as it relates to armed conflict.

Wartime Mimicry Through History

Although this book focuses on modern civil wars, the phenomenon we are dealing with has ancient roots. Sayings linking conflict and deception span all of literary history, from the Epic of Gilgamesh (twelfth century BC): 'mankind is deceptive and will deceive you' (Kovacs 1989: 104), via Sun Tzu (sixth century BC): 'all warfare is based on deception' (Sun-Tzu 2004: 31), and the Prophet Muhammad (seventh century AD): 'war is deceit' (Hayward 2017: 7–8), to William Taverner's (eighteenth century AD): 'all advantages are fair in love and war' (Taverner 2018: 57). Trickery features in most tales of war, including humanity's oldest. In *The Taking of Jaffa* (Petrie 2013: 1–12), an ancient Egyptian story recorded in the fifteenth century BC, General Djehuty places soldiers inside sacks of grain and leaves them at the city gate. Jaffa's residents mistake the sacks for a tribute offering and bring them inside, allowing the Egyptians to capture the city. (The similar but more famous story of the Trojan horse features in *The Odyssey*, written some seven centuries later.) The historical record contains thousands of anecdotes about deception in war.

So numerous were these stories already in antiquity that books were written to make sense of them. The Roman engineer Frontinus (first century AD) and the Greek writer Polyaenus (second century AD) both wrote treatises titled *Stratagems of War* containing a total of around 1300 examples of *ruses de guerre* deployed by the Greeks, the Romans, and their enemies (Bennett 1925; Krentz and Wheeler 1994). The historicity of the vignettes is questionable, but they testify to a broader interest and apprehension regarding wartime deception. Indeed, these two works are but the surviving parts of what is believed to have been an entire genre in classical literature (Wheeler 1988: ix).

The topic of deceptive tactics also features prominently in many manuals of war produced across the centuries.[11] Some 2600 years ago, Sun Tzu devoted two chapters of his famous book *The Art of War* to deception. Several medieval Arabic works, such as al-Harawi's *Advice on Stratagems of War* (twelfth century AD) and al-Ansari's *Dispelling of Woes in the Management of Wars* (fourteenth century AD), include chapters on deception (Scanlon 1961; Sourdel-Thomine 1962). The pattern continues up to the nineteenth and twentieth centuries, with publications such as *The Official CIA Manual of Trickery and Deception*, a 1950s field manual for American spies (Melton and Wallace 2010). Modern academics have also paid attention to the historical use of deception and have documented its extensive use in past conflicts (Dewar 1989; Holt 2004; Latimer 2001; Muñoz-Rengel 2022; Sheldon 2003; Whaley 2007). Of particular note is a recent book by the historian John Titterton (2022) on deception in medieval warfare, which documents hundreds of examples from eleventh- to fourteenth-century Europe.

Many of the historical treatises reveal a certain normative appreciation for deception in war. Sometimes, the same societies that meted out severe punishments for theft and dishonesty in economic transactions would

---

[11] Not all manuals of war speak of deception, and there are some interesting exceptions. For example, the main Byzantine war manuals such as the *Strategicon* of Maurice (sixth or ninth century) or Leo's *Tactica* (ninth–tenth century) do not seem to address it.

applaud generals who lied and deceived their way to victory. The ancient Greeks, in particular, in what might be the dark side of their admiration for philosophical and cognitive acumen, respected cunning enemies who knew how to play dirty tricks. For example, in the beginning of book seven of his *Stratagems*, Polyaenus writes that 'even the minds of barbarians are capable of military stratagems, deceptions, and tricks. And therefore you will see that you yourselves should not hold them in too great contempt, and your generals must be similarly cautious'. To be sure, not all strategists were infatuated with clever ruses. The most prominent exception is probably Clausewitz, who, in his nineteenth-century classic *On War*, dismissed trickery and deception as a distraction from the more consequential work of strategic disposition of forces, noting that 'a correct and penetrating eye is a more necessary and more useful quality for a General than craftiness' (von Clausewitz 1997: 175). The twentieth century would see more normative pushback, notably with the banning of certain deceptive tactics in the Geneva conventions, such as the misuse of ambulances, hospitals, places of worship, and personnel categories associated with these institutions. These conflicting views on deception reflect what Wheeler (1988 pp. xiii–xiv) has referred to as the Achilles ethos and the Odysseus ethos in the Western military tradition, with the first valuing honesty and chivalry and the second appreciating cunning and deceit. There is arguably an echo of this tension in contemporary Western society's simultaneous appreciation of good spycraft and advocacy for international law.

However, neither early chroniclers nor modern scholars appear to have conceived of identity mimicry as a distinct type of deception. Some works devote special sections to disguises or spying, but most often the examples of identity mimicry are buried within generic historical prose or compilations of *ruses de guerre*. This is surprising when we consider that early biologists, such as Henry W. Bates and Fritz Müller, observed and studied many varieties of mimicry in other animals since the middle of the nineteenth century, though perhaps the reason is that in biology the focus has been on mimicry between rather than within species.[12]

This is not to say that the recorded examples of human mimicry are few and far between. Where there are war stories, there is usually identity mimicry. Some of the earliest accounts stem from ancient China. In the sixth century BC, Sun Tzu described how a team of scouts working for Emperor T'ai Tsu had been able to infiltrate an enemy camp and steal their password by dressing up as enemy soldiers (Sun-Tzu 2004: 126). A century or two later, the essay

---

[12] See 'Mimicry (biology)' on www.britannica.com.

'Thirty-six Stratagems' included an approach called 'Shed your skin like the golden cicada', a reference to the tactic of escaping from a difficult situation by pretending to be someone else (Verstappen 1999: 99). A subsequent account relates how Liu Bang (later known as Emperor Gaozu of Han) escaped the siege of Xingyang in 204 BC by having one of his generals pretend to be him and ride out the city offering to surrender, thereby distracting the besiegers, allowing the real Liu Bang to slip away (Verstappen 1999: 100).

Frontinus and Polyaenus' *Stratagems* also contain a wealth of mimicry stories, some of which go back to the sixth and seventh centuries BC if we are to believe the compilers.[13] In our rough estimation, around ten to twenty per cent of the 1300 or so stories in Frontinus and Polyaenus involve mimicry. The majority represent offensive mimicry, typically the disguising of one's own soldiers—as enemy personnel, as prisoners, as women, or some other inoffensive category—in order to attack by surprise or access a city. We read, for example, that the Arcadians, 'when besieging a stronghold of the Messenians, fabricated certain weapons to resemble those of the enemy. Then, at the time when they learned that another force was to relieve the first, they dressed themselves in the uniform of those who were expected, and being admitted as comrades in consequence of this confusion, they secured possession of the place and wrought havoc among the foe' (Bennett 1925: 209). However, there are also examples of mimicry used defensively. For example:

> Artemisia [I of Caria] always chose a long ship, and carried on board with her Greek, as well as barbarian, colours. When she chased a Greek ship, she hoisted the barbarian colours; but when she was chased by a Greek ship, she hoisted the Greek colours; so that the enemy might mistake her for a Greek, and give up the pursuit (Shepherd 1793: 354).

Many stories involve groups mimicking other groups, both by a set of individuals using another group's members observable features or by the display of another group's collective insignia. There are also multiple examples of individual mimicry. We have individuals mimicking specific individuals for offensive purposes; for example, 'Timarchus, the Aetolian, having killed Charmades, general of King Ptolemy, arrayed himself in Macedonian fashion in the cloak and casque of the slain commander. Through this disguise he

---

[13] Frontinus is probably the earliest source where mimicry episodes are described in detail. However, wartime mimicry must have occurred earlier, if only in the context of espionage, which is documented back to at least the second millennium BC, notably in ancient Egypt (Sheldon 2003).

was admitted as Charmades into the harbour of the Sanii and secured possession of it' (Bennett 1925: 213). We also have individuals mimicking generic
figures for offensive advantage, as in the case of Zopyrus who:

> mangled his face horribly and fled to the enemy, in the guise of a deserter. He pre-
> tended that they had been cruelly treated by Dareius, and the Babylonians believed
> his complaints, which were so clearly supported by his appearance. They took him
> under their protection, and their confidence in him increased by degrees, until at
> last they entrusted him with the government of the city. After he had been invested
> with this power, he soon found the means to throw open the gates by night, and
> allowed Dareius to take possession of Babylon (Bennett 1925: 215).

And there are multiple stories of leaders escaping sieges and attacks in
disguise. For example, 'After Athens was captured by Demetrius, Lachares
slipped out through a little gate, in a slave's clothes, with his face blackened,
and a basket of money covered with dung on his arm' (Shepherd 1793: 100).
There are also some peculiar episodes that are not easy to classify, such as
that of King Codrus of Athens, who, in order to usher in a prophecy which
said Athens would triumph if Codrus was killed by a Peloponnesian, dressed
up as a woodsman and picked a fight with some Peloponnesian soldiers
in order to get himself killed and have the prophecy come true (Shepherd
1793: 17).

Interestingly, Frontinus and Polyaenus describe almost exclusively successful mimicry attempts. We rarely hear about people failing and getting caught,
perhaps because they did not live to tell their tale or because the chroniclers
believed unsuccessful attempts were not as exciting. There are exceptions,
however; for example:

> When Memnon advanced against Cyzicus, he put a Macedonian cap upon his head,
> and made all his army do the same. The generals of Cyzicus, observing their
> appearance from the walls, supposed that Chalcus the Macedonian, their friend
> and ally, was marching to their assistance with a body of troops; and opened their
> gates to receive him. However, they discovered their error just soon enough to
> correct it and shut their gates against him; Memnon had to content himself with
> ravaging their country (Shepherd 1793: 228).

In mimicry accounts from the medieval period, we encounter many of the
same types of scenarios (Titterton 2022). People disguise to attack: in 1070,
Hereward the Wake disguises himself as a fisherman to enter a Norman
camp; in 1116, Louis VI of France disguises his men to attack the town

of Gasny in Normandy; and in 1192, Richard I, on crusade in Palestine, sends out spies disguised as Bedouin. Around 1013, Hugh, Count of Maine, flees to Le Mans disguised as a shepherd; in 1098, the ruler of Antioch flees the city disguised as a beggar; in 1116 and in 1322, Thomas, Earl of Lancaster, and his men flee battle disguised as beggars. Titterton lists 427 deception events between 1000 and 1320, about fifty of which involved a form of mimicry. This number is our estimate, because Titterton does not distinguish between mimicry and other forms of deception. He sorts events into ten categories, two of which (disguise and spying) appear to correspond to mimicry, but the examples include many instances of false pretense.

One of the most striking features of the medieval episodes is the relative ease with which prominent leaders, even high royalty, could mimic and be mimicked. This was symptomatic of a systemic problem of identity verification in the pre-modern era. As Eliav-Feldon (2012: 1) notes:

> Early modern Europe was teeming with impostors. Men and women from all walks of life were inventing, fabricating and disguising themselves, lying about who they were or pretending to be someone they were not. As a result, authorities, both religious and secular, were frantically creating new means for ascertaining each person's identity.

In addition to the many stories of leaders escaping dangerous situations by pretending to be nobodies, there are several cases of nobodies pretending to be very prominent people and nearly succeeding in assuming their positions, because relatively few people knew what the prominent people actually looked like (Kalinin 2017). One of the most famous of these cases is that of 'the false Olaf' in which a mysterious man turned up in Denmark in 1402 pretending to be Prince Olaf II of Norway—who had died fifteen years earlier in murky circumstances—and claiming the Danish throne. He succeeded in acquiring a substantial following, but was ultimately dismissed as an impostor and executed. There is some evidence that the impostor was part of a German ploy to foment internal unrest in Denmark in preparation for an invasion (Cole 2022). Unfortunately, no accounts survive of the means by which the Danish authorities sought to establish whether the new Olaf was genuine or not. This is a general shortcoming of the historical accounts of mimicry; they are short on the details of the signalling game surrounding each mimicry attempt.

The medieval and early modern periods also feature a curious type of wartime mimicry that is neither offensive or defensive, but rather designed to bypass prohibitions in one's own camp. This is the phenomenon of women pretending to be men (and occasionally underage men pretending to be older) in order to be allowed to fight (Dugaw 1996; Hall 1993; Verbruggen 2006; Wheelwright 1989). The Islamic tradition includes the famous example of Khawla bint al-Azwar, who, according to some sources, fought several battles in the Muslim conquest of the Levant in the 630s AD dressed as a man (Qazi 2011). Several similar cases were recorded in Europe in the Middle Ages, but the phenomenon really proliferated, at least in terms of reported cases, in the eighteenth and nineteenth centuries. In 1762, according to De Pauw (2000: 105), 'an Englishman joked that so many disguised women were serving in the army that they ought to have their own regiments'. Elizabeth Leonard (1999: 165, 310) estimates that somewhere between 500 and 1000 women fought in the American Civil War in the 1860s dressed as men. Here the dupes have a material advantage to be duped, so they may choose to turn a blind eye and pretend to be duped rather than violate the norm.

An extraordinary case of this sort, recounted by Bryan Mark Riggs (2002), is that of German Jews who fought with the German army in the Second World War. Rigg estimates that as many as 150,000 men were in this position. The army was not exactly duped, but self-interestedly lenient allowing *Mischlinge*, men of partial Jewish ancestry, to serve in the army. In some cases, exemptions were handed out to men and also their families, some signed by Hitler himself, and Aryan purity was imposed by fiat. Later in the war, however, this policy was tightened, and Nazi ideology won over military logic, despite the growing German need for military personnel.

As we move closer to the modern era, wars become substantially better documented, and for major events like the two world wars, it no longer makes sense to catalogue individual mimicry episodes, as there were simply too many of them. In most of German-occupied Europe in the Second World War, for example, there were underground resistance movements whose members likely engaged in various kinds of mimicry on an almost daily basis. It is also in the twentieth century that we start getting accounts of the nitty-gritty of signal interpretation in mimicry games. This comes out clearly in Finkel (2017), which documents many nerve-wracking episodes in which Jews tried to pass off as non-Jews to escape Nazi persecution. The twentieth-century sources also witness the professionalization of state-run mimicry and counter-mimicry in war. It is in this period that states develop large intelligence services and as well as formal military strategies of deception, such as the Soviet doctrine of *maskirovka* (Holt 2004; Rankin 2009).

The historical sources also contain many examples of mimicry counter-measures. The ancients were not gullible; even the oldest mimicry accounts have the dupes apprehensive. When the biblical Isaac is approached by Jacob pretending to be Esau, Isaac first asks him suspiciously how he had made it back from his hunting trip so quickly. He then commands: 'Come near so I can touch you, my son, to know whether you really are my son Esau or not'. In the end Isaac was tricked, but it was not for lack of precaution.

The Bible also features one of history's most famous screening devices, namely, that of the Shibboleth. Judges 12: 5–6 relate how the men of Gilead used a dialect difference to distinguish between real Gileadites and Ephraimites pretending to be Gileadites:

> The Gileadites captured the fords of the Jordan leading to Ephraim, and whenever a survivor of Ephraim said, 'Let me cross over', the men of Gilead asked him, 'Are you an Ephraimite?' If he replied, 'No', they said, 'All right, say "Shibboleth".' If he said, 'Sibboleth', because he could not pronounce the word correctly, they seized him and killed him at the fords of the Jordan. Forty-two thousand Ephraimites were killed at that time.

The word Shibboleth has since taken on the generic meaning of any word whose pronunciation can be used to tell insiders from outsiders. The method exploits the fact that nuances of pronunciation are often acquired in childhood and are thus unconscious or in any case hard to suppress at will. Shibboleths have remained a powerful mimicry screening device to this day.[14] For example, in the war in Ukraine in 2022, Ukrainians reportedly used the word *palianytsia* (a type of Ukrainian bread) as a test to weed out Russian agents (Lim 2022).

Early writers such as Polyaenus and Frontinus included ploys to weed out spies, which suggests they considered cunning countermimicry as valiant as any other *ruse de guerre*. Polyaenus, for example, describes how:

> Chares, who suspected that the enemy had spies in his camp, placed a strong guard outside the trenches, and ordered every man to question his neighbour, and not to part till each had told the other, who he was, and to what company, and band, he belonged. By this device the spies were revealed and caught: because they were unable to tell either their company, band, comrade, or the password (Shepherd 1793: 130).

---

[14] See the *Wikipedia* page for 'Shibboleth' for a long list of historical examples.

He also writes that:

> when Theognis suspected that spies had infiltrated into the camp, he posted guards on the outside of the trenches, and then ordered every man to take his station by his own weapons. In consequence of this order, the spies became easy to distinguish; either because they moved away, or because they had no weapons by which to post themselves (Shepherd 1793: 218).

This brief review of wartime mimicry through history raises the question of how much has changed and how much stays the same. On the one hand, there are striking continuities, such as the use of Shibboleths for at least three thousand years. On the other hand, the change of mores and technology change have made some early mimicry tactics irrelevant. Today, it would make little sense for a large military unit to march towards an enemy base dressed as musicians in the hope of slipping inside (as the ancients did on several occasions according to Polyaenus), because musical bands do not often march down streets in this era and in any case the defenders would spot them with binoculars ahead of time or screen them with metal detectors at the gate. Similarly, it is almost inconceivable that a person could turn up in a European capital today claiming to be some long-lost head of state the way the False Olaf did in 1402. These changes illustrate an important aspect of deceptive mimicry, namely, that the devilish details that make or break it vary by context.

Mimicry in Modern Civil Wars

What forms, then, might mimicry take in contemporary conflicts? We can formulate some broad expectations by combining what we have just said about mimicry and what we know about modern civil wars. The archetypal civil war will have three ideal-type protagonists: rebels, government forces, and civilians. In principle, all of them can make use of deceptive mimicry, all can serve as models to the mimicking agents, and all can be the dupe, the intended victim of a mimicry act. Given the creativity with which past conflict actors have deployed mimicry, and given that some conflicts are more complex than the classic 'single rebel group versus government' constellation, it is reasonable to expect a broad range of mimicry scenarios. For simplicity's sake, we can group them into four categories based on the type of model involved.

One set of episodes may involve individual mimicry of specific individuals—offensively, for example, by pretending to be a VIP to access sensitive premises in connection with an operation; defensively, for example, in the form of a military leader using doubles. In this case, the model himself recruits the mimic to avoid exposing himself to dangerous situations and letting the mimic do that instead (as the following joke conveys: during the first Gulf War, an Iraqi general summoned Saddam Hussein's doubles and told them: 'the good news is that Saddam is alive, the bad news is that he lost an arm').

Another category of likely scenarios involves individual mimicry of generic representatives of other categories of people. This is likely one of the most widespread types of conflict mimicry, as it can be used by individual operatives both to infiltrate enemy circles and to extract themselves from difficult situations. In Thomas Hegghammer's contribution (Chapter 3), the key source of trust problems on jihadi web forums is the fear that some members may be police infiltrators. Rebels may also mimic civilians for defensive purposes. A recent example is that of the many ISIS fighters in Syria and Iraq who, during the fall of the Caliphate around 2017, dressed as women to escape capture at the hand of security forces. In Chapter 6, David Laitin looks for variants of this mimicry type in the context of the Rwandan genocide in 1994.

A third set of mimicry episodes involves groups mimicking other groups. Again, the objective might be offensive, defensive, or reputational. A rebel patrol might pretend to be a government unit to get within striking distance of its target. There have been several notable cases in Afghanistan of groups of Taliban fighters pretending to be Afghan security patrols so as to gain access to the latter's bases. Conversely, many recent conflicts have seen rebel groups pretending to be humanitarian organizations in order to exploit the protections afforded such groups under international law to move personnel, goods, or money undetected. This is arguably how terrorism financing has often worked historically. The reputational variant of group mimicry is probably also widespread, for example, in the form of the false flag operation, in which a group does something in the name of another group to tarnish the latter's reputation. We have also seen unaffiliated rebels carry out attacks in the name of a more established group to piggyback on the latter's reputation and increase the psychological impact of an attack; think amateur jihadis claiming to be al-Qaida or ISIS when there was no link in reality. It can also include the situations where established groups operate in the name of fictitious groups so as to avoid incurring the political cost of controversial attacks. In such cases, the model group is specific but invented. Chapter 5 by Valeria Pizzini-Gambetta and Diego Gambetta goes deep on this, looking at the dynamics triggered by false claims of responsibility for terrorist attacks by far-left and far-right groups in Italy in the late 1970s.

The fourth scenario type involves groups mimicking generic collectives. The typical example would be groups of fighters pretending to be civilians to avoid raising suspicion as they deploy an attack or to avoid capture when they are on the defensive. In Chapter 8, Stephen Holmes surveys this and other types of mimicry observed in the Iraqi war theatre in the years following the US invasion in 2003.

The picture gets more complicated if we throw criminal actors into the mix. In Colombia there are reported instances of both criminal kidnappers passing off as insurgents to increase the credibility of their blackmail, and, conversely, insurgents passing off as criminals when kidnapping or robbing a bank often in order not to tarnish their political image.[15] Conceivably we can also imagine cases of double mimicry; there were cases in Iraq in the 2000s where real rogue policemen kept their police uniforms while carrying out violent acts, counting on the fact that people would believe they were insurgents disguised as policemen. In Chapter 7, Pavan Mamidi looks at how the Indian left-wing guerrilla group known as the Naxalites protected their brand against criminal exploitation of their signature through mimicry.

Given what we know about the game dimension of mimicry, we should also expect to see systematic countermimicry efforts in civil wars. Anecdotally there are indeed ample indications of this, from rebel groups issuing ID cards, via militants vetting new recruits through extensive questioning, to the widespread practice of torturing and executing suspected spies and infiltrators. In Chapter 4, Heather Hamill examines the use of codewords as an identity signal and anti-mimicry device in the Northern Ireland conflict in the 1970s.

In 1932 the American physiologist Walter Cannon coined the term 'fight-or-flight response' to describe the typical animal reaction to lethal threats. Fight and flight remain fundamental alternatives in human conflict, but both

---

[15] The best-known case of the latter kind is that of journalist Diana Turbay, daughter of former Colombia president César Turbay (1978–1982), who was kidnapped by Pablo Escobar, head of the so-called Medellin cartel, as part of a series of high-level kidnappings to put pressure on the government not to approve his extradition to the US. The gang passed themselves off as members of the ELN left-wing guerrilla group and convince her to go with them by claiming an ELN commander (El cure / The priest) wanted to have an interview with her to discuss peace prospects, as she has done already with a commander of another guerrilla, the M19. The story is told by Gabriel Garcia Márquez (1998).

are often supported by the strategic use of identity signalling, as the contributions in this volume will show, by taking us deep into the complex world of wartime mimicry. Taken together, the chapters cover a wide but non-exhaustive range of dynamics, let alone cases. This is because we chose to let the authors select and explore natural case studies about which they are knowledgeable instead of assigning them specific types or cases of mimicry to study. There is room, in other words, for more work on this topic, and we hope this volume will constitute a thought-provoking start.

References

Bell, J. Bowyer. 1982. *Cheating: Deception in War & Magic, Games & Sports, Sex & Religion, Business & Con Games, Politics & Espionage, Art & Science*. New York: St Martin's Press.

Bell, J. Bowyer and Barton Whaley. 1991. *Cheating and Deception*. New Brunswick, NJ: Routledge.

Bennett, Charles E. 1925. *Frontinus: Stratagems. Aqueducts of Rome*. Loeb Classical Library No. 174. Cambridge, MA: Harvard University Press.

Braun, Robert. 2019. *Protectors of Pluralism: Religious Minorities and the Rescue of Jews in the Low Countries during the Holocaust*. Cambridge: Cambridge University Press.

Caddell, Joseph William. 2004. *Deception 101: Primer on Deception*. Carlisle, PA: US Army War College.

Carpenter, G. D. Hale and E. B. Ford. 1933. *Mimicry*. London: Methuen & Co.

Clark, Robert M. and William L. Mitchell. 2018. *Deception: Counterdeception and Counterintelligence*. London: CQ Press.

Clausewitz, Carl von. 1997. *On War*. Ware: Wordsworth.

Cole, Richard. 2022. 'The False King Olaf, Queen Margaret, and the Prussian Hansa'. *Viking and Medieval Scandinavia* 18: 83–111.

De Pauw, Linda G. D. 2000. *Battle Cries and Lullabies: Women in War from Prehistory to the Present*. Norman, OK: University of Oklahoma Press.

Dewar, Michael. 1989. *The Art of Deception in Warfare*. Newton Abbot: David & Charles.

Donath, Judith S. 1998. 'Identity and Deception in the Virtual Community'. In *Communities in Cyberspace*, edited by Peter Kollock and Marc Smith, 37–68. Routledge: London.

Dugaw, Dianne. 1996. *Warrior Women and Popular Balladry, 1650–1850*. Chicago, IL: University of Chicago Press.

Eliav-Feldon, Miriam. 2012. *Renaissance Impostors and Proofs of Identity*. London: Palgrave Macmillan.

Finkel, Evgeny. 2017. *Ordinary Jews: Choice and Survival during the Holocaust*. Princeton, NJ: Princeton University Press.

Forbes, Peter. 2009. *Dazzled and Deceived: Mimicry and Camouflage*. New Haven, CT: Yale University Press.

Gambetta, Diego. 2005. 'Deceptive Mimicry in Humans'. In *Perspectives on Imitation: From Neuroscience to Social Science*. Volume 2: *Imitation, Human Development and Culture*, edited by Susan L. Hurley and Nick Chater, 221–242. Cambridge, MA: MIT Press.

Gambetta, Diego. 2009a. *Codes of the Underworld: How Criminals Communicate*. Princeton, NJ: Princeton University Press.

Gambetta, Diego. 2009b. 'Signalling'. In *The Oxford Handbook of Analytical Sociology*, edited by Peter Hedström and Peter S. Bearman, 168–194. Oxford: Oxford University Press.

Gambetta, Diego and Heather Hamill. 2005. *Streetwise: How Taxi Drivers Establish their Customers' Trustworthiness*. New York: Russell Sage.

Glantz, David M. 1989. *Soviet Military Deception in the Second World War*. London: Routledge.

Hall, Richard. 1993. *Patriots in Disguise: Women Warriors of the Civil War*. New York Marlowe & Co.

Handel, Michael I. 1982. 'Intelligence and Deception'. *Journal of Strategic Studies* 5 (1): 122–154.

Hayward, Joel. 2017. *War is Deceit: An Analysis of a Contentious Hadith on the Morality of Military Deception*. Amman: The Royal Islamic Strategic Studies Centre.

Holt, Thaddeus. 2004. *The Deceivers: Allied Military Deception in the Second World War*. London: Weidenfeld & Nicolson.

Kalinin, Ilya. 2017. 'The Figure of the Impostor to the Throne in Russian Political Culture: Between Sacralization and Mimesis'. *Σημειωτκή—Sign Systems Studies* 45 (3–4): 284–301.

Kovacs, Maureen G. 1989. *The Epic of Gilgamesh*. Stanford, CA: Stanford University Press.

Krentz, Peter and Everett L. Wheeler. 1994. *Polyaenus Stratagems of War*. Chicago, IL: Ares.

Latimer, Jon. 2001. *Deception in War: The Art of the Bluff, the Value of Deceit, and the Most Thrilling Episodes of Cunning in Military History, from the Trojan Horse to the Gulf War*. Woodstock, NY: Overlook Press.

Leonard, Elizabeth D. 1999. *All the Daring of the Soldier: Women of the Civil War Armies*. New York: W. W. Norton & Company.

Lim, Lisa. 2022. 'The Ukrainian word that uncovers Russian soldiers, and others that, mispronounced, could have meant your death'. *South China Morning Post*, 2 July 2023.

Martin, Clancy (ed.). 2009. *The Philosophy of Deception*. Oxford: Oxford University Press.

Márquez, Gabriel García. 1998. *News of a Kidnapping*. London: Penguin.

Melton, H. Keith and Robert Wallace. 2010. *The Official CIA Manual of Trickery and Deception*. Repr. edn. New York: William Morrow.

Muñoz-Rengel, Juan J. 2022. *A History of Lying*. Cambridge: Polity.

Petrie, W. M. Finders. 2013. *Egyptian Tales, Translated from the Papyri First series, IVth to XIIth dynasty*. Cambridge: Cambridge University Press.

Qazi, Farhana. 2011. 'The Mujahidaat: Tracing the Early Female Warriors of Islam'. In *Women, Gender, and Terrorism*, edited by Laura Sjoberg and Caron E. Gentry, 29–56. Athens, GA: University of Georgia Press.

Rankin, Nicholas. 2009. *A Genius for Deception: How Cunning Helped the British Win Two World Wars*. Oxford: Oxford University Press.

Riggs, Bryan M. 2002. *Hitler's Jewish Soldiers: The Untold Story of Nazi Racial Laws and Men of Jewish Descent in the German Military*. Lawrence, KS: University of Kansas Press.

Scanlon, George T. 1961. *A Muslim Manual Of War, Being Tafrīj Al-Kurūb Fi Tadbīr Al-Hurūb*. Cairo: American University in Cairo Press.

Sheldon, R. M. 2003. *Espionage in the Ancient World: An Annotated Bibliography of Books and Articles in Western Languages*. Jefferson, NC: McFarland & Company.

Shepherd, Richard. 1793. *Polyaenus's Stratagems Of War*. London: George Nicol.

Sourdel-Thomine, Janine. 1962. 'Les Conseils du Sayh al-Harawi à un Prince Ayyubide'. *Bulletin d'Études Orientales* 17: 205–268.

Sun, Lixing. 2023. *The Liars of Nature and the Nature of Liars: Cheating and Deception in the Living World*. Princeton, NJ: Princeton University Press.

Sun-Tzu. 2004. *The Art Of War*. Project Gutenberg (https://www.gutenberg.org).

Taverner, William. 2018. *The Artful Husband. A Comedy. As it is Acted at the Theatre in Lincoln's-Inn-Fields*. Farmington Hills, MI: Gale ECCO.

Titterton, James. 2022. *Deception in Medieval Warfare: Trickery and Cunning in the Central Middle Ages*. Martlesham: Boydell & Brewer.

Verbruggen, Jan F. 2006. 'Women in Medieval Armies'. In *Journal of Medieval Military History IV*, edited by Clifford J. Rogers, John France, and Kelly Devries, 119–136. Martlesham: Boydell & Brewer.

Verstappen, Stefan H. 1999. *The Thirty-six Strategies Of Ancient China*. San Francisco, CA: China Books & Periodicals.

Wickler, Wolfgang. 1968. *Mimicry in Plants and Animals*. New York: McGraw-Hill.

Whaley, Barton. 2007. *Stratagem: Deception and Surprise in War.* Boston, MA: Artech House.

Wheeler, Everett. L. 1988. *Stratagem and the Vocabulary of Military Trickery.* Leiden: Brill.

Wheelwright, Julie. 1989. *Amazons and Military Maids: Women who Dressed as Men in the Pursuit of Life, Liberty and Happiness.* London: Pandora Press.

2
Strategic Dynamics of Social Mimicry

James D. Fearon

Gambetta (2005) describes situations of social mimicry in terms of a 'mimic' who chooses from a set of signals in hopes of deceiving a 'dupe' into thinking that he or she possesses some unobservable property k. In Gambetta's paradigmatic cases, the property k is membership in a social category (an 'identity'), but in some applications it might not be. Examples might include:

- a minor using a fake driving licence to buy alcohol;
- a Tutsi pretending to be a Hutu in an attempt to escape the 1994 genocide in Rwanda (Laitin, in this volume);
- an IRA operative pretending to be a non-combatant in encounters with the authorities in Northern Ireland (Heather Hamill, in this volume);
- a would-be terrorist attempting to board an airplane in order to hijack or blow it up;
- a taxpayer misrepresenting her annual income on tax forms.

I describe below a simple game-theoretic model of social mimicry, which proves to have the underlying structure of betting in a card game like poker. Mimicry is analogous to bluffing. It may be useful to develop the analogy formally in order to make precise the nature of the strategic interaction in social mimicry and to produce comparative statics results interpretable for this class of situations.

Social mimicry is a strategic problem in the classic sense that optimal behaviour for the mimic depends on what the target (Gambetta's 'dupe') is expected to do, and vice versa. If targets always investigate and have a high enough probability of detecting mimics, then it may not be worthwhile for mimics to make the attempt. But if attempted mimicry is quite rare, then targets have little incentive to make the effort to try to detect it.

These observations already suggest that in an equilibrium state of affairs, the expected payoff from attempting to mimic must be about the same as the expected payoff of not mimicking. If mimicry is better than non-mimicry,

James D. Fearon, *Strategic Dynamics of Social Mimicry*. In: *Fight, Flight, Mimic*. Edited by: Diego Gambetta and Thomas Hegghammer, Oxford University Press. © James D. Fearon (2024). DOI: 10.1093/9780191802454.003.0002

more mimicry will be attempted and detection efforts will increase. If mimicry is worse on average than non-mimicry, then fewer will try it and detection efforts will decrease, restoring the incentive for more mimicry. Likewise, in equilibrium the rate of mimicry must be such that the expected payoff to a target of attempting to detect must be about the same as the payoff for not bothering. If detection efforts fall below this rate, then more mimicry is attempted, raising the payoff to trying to detect, and vice versa if detection efforts are too intense.

As shown formally below, an interesting implication is that if detection efforts are made sometimes but not always in equilibrium, then the rates at which mimicry is attempted and succeeds depend on how much the *target* (or 'dupe') cares about stopping it versus the costs of doing so, and not at all on how much mimics want to pass or their costs for getting caught trying. When the value of detecting mimics (or equivalently, the costs of failing to detect) is high, targets will attempt detection for sure unless the rate of attempts is low. Thus in equilibrium attempts will be rare.

This observation might help explain the possible fact that attempts to pass in genocides are surprisingly infrequent (Laitin, in this volume). More generally, the interesting result is that there are conditions such that the higher the *stakes* at issue—it is reasonable to think that in many cases the value of passing for the mimics is related to the value of detection for the targets—the *lower* the equilibrium rate of attempted mimicry, the greater the equilibrium detection efforts, and the lower the rate of successful passing.

The analysis produces a large set of comparative statics, which could be useful for empirical assessments in particular areas. There are quite a few potential variables of interest here, including:

- the percentage of non-k types (potential mimics) in the population
- the percentage of non-k types who try to pass
- the percentage of non-k types who successfully pass
- the percentage of non-k types who are caught trying to pass
- the percentage of people in the population exhibiting k signals who are mimics
- the frequency of detection efforts
- the probability that a mimic passes.

Depending on the specific empirical example, some of these may be known or observable, while others are certainly not. For example, we might know the percentage of fifteen to twenty year olds in the population and the number caught trying to buy alcohol with a fake ID card, but not the number with fake ID cards. The formal results can establish expectations about how these various quantities might be expected to be related to each other, and how they would vary with factors like the stakes, the probability of detection conditional on inspection, the cost for a mimic of being detected, the share of non-k types in the population, and so on.

The next two sections describe and analyse the model. I then consider two extensions. First, would more realistic evolutionary or adaptive dynamics yield convergence on the model's equilibrium rates of mimicry? Second, under what conditions would the 'dupe', who is often a government or other organization acting through an agent, want to develop institutions that commit its agents to investigate every potential mimic? The fifth section then considers specific empirical illustrations in light of the model results. In the conclusion, I compare the strategic problem of mimicry to that of signalling private information in the classic Spence (1973) model, and more generally to the idea of political and economic institutions as solutions to problems of mechanism design. The core function of a great many political and economic institutions is to enable people possessing an unobservable property k (in Gambetta's terms) to credibly reveal this, distinguishing themselves from and so deterring potential mimics. Mimicry may be quite rare in a well-functioning advanced industrial economy and society, but this is because the institutions that make up the economy and society are designed to discourage mimicry. Similarly, following on the results concerning institutional commitments to investigate potential mimics, we should expect that in societies where ethnicity or religion are highly politicized, dominant groups will develop institutions that dedicate resources to screening for mimics, so keeping equilibrium rates low despite strong incentives.

The Model

We consider the interaction between a person who may or may not want to pass herself off as possessing the valuable property A (which could be membership in some social category), and a potential target or 'dupe', which I will call the government, or G for short. The game begins with Nature drawing the person who will interact with G from society, represented as a continuum of agents. A share $1 - \beta$ of society possesses the desirable characteristic A, and β does not. Those who possess the characteristic A will be called As, and those who do not will be called Bs. A and B could thus refer to two social groups like Hutus and Tutsis in Rwanda before 1995. A B who tries to pass as an A will be called a 'mimic'.

Bs differ among themselves in how costly they would find it to try to pass as an A. An individual B's cost of trying to pass is $e \geq 0$, where in the population of Bs e is distributed by a smooth cumulative distribution function F. These costs could be material, if mimicry requires acquiring certain goods, skills, or changing one's appearance, or psychological, if associated with costs for abandoning one's identity as a B or losing social connections.[1] After Nature draws the person to interact with the potential dupe, the person chooses whether to try to pass. A types always try to pass, at zero cost. Bs who decide not to try to pass get their value for the status quo, which is normalized to be zero. The government (or government agent) then observes whether the person is emitting A signals or not, and chooses whether to investigate if the person is a mimic. This would involve any of the various ways that people have of detecting deception, as appropriate for the specific situation. For instance, they might ask for and inspect an ID card in the underage-drinking example, or interrogate the suspect in the IRA example. Assume that investigating has a fixed cost $d > 0$ (d for 'detect').

If the government chooses not to investigate, then G realizes a payoff of zero if the person was in fact an A, and suffers a loss $l > 0$ if she was a mimic. For example, l might be the expected cost of fines and loss of business if the police discover that the liquor store sold to an underage buyer. In the IRA example, the cost of failing to detect a mimic is the political and material damage that follows (in expectation) from not jailing an IRA operative.

If the government chooses to investigate, then I assume for convenience that it correctly identifies an A as an A type for sure. However, it correctly identifies a mimic as a mimic only with probability $1 - p$, so that $p \geq 0$ is the probability that a mimic succeeds in passing as an A if investigated.

A mimic who passes gets the benefit $v > 0$. A mimic who is detected pays a cost $c > 0$. Thus if the government chooses to investigate and the person is in fact a mimic, the mimic's expected payoff is $pv - (1 - p)c - e$. If G chooses not to investigate, the mimic gets $v - e$. Likewise, the government gets an expected payoff of $-pl - d$ for investigating when it in fact faces a mimic. If the government in fact faces an A, then investigation just wastes the effort cost d (yielding payoff for G of $-d$). Table 2.1 summarizes the payoffs in different scenarios and outcomes.

For many cases, it is natural to suppose that the mimic's value for passing is related to the government's cost of failing to detect a mimic. For example, if

---

[1] An alternative formulation that is more natural for some examples is to make the cost of mimicking fixed, but assume that Bs vary in their value for passing as an A. That is, we would take e as fixed and let the benefit of passing (v below) be distributed according to F. The results in this variant are essentially identical.

Table 2.1 Payoffs from an encounter

Type of person	G's action	Mimic's payoff	G's payoff
Mimic	Investigates	$pv - (1 - p) c - e$	$-pl - d$
	Does not investigate	$v - e$	$-l$
A Type	Investigates		$-d$
	Does not investigate		0

Source: created by author

the As are allowed access to some desirable goods that are denied to Bs—such as government jobs, more positions in the university, social distinction—then more for the Bs can mean less for As. In the case of a pure taste for discrimination, it may be that the more the A types hate the Bs, the greater the value of passing as an A. So in the comparative statics below we will sometimes consider what happens as v and l are varied together (e.g. $v = l$), and in this case we will refer to the *stakes* at issue in mimicry.

Analysis

The game is easily solved for a perfect Bayesian equilibrium in which G and every type of B are doing the best they can given the other players' strategies, and in which the government's beliefs about the likelihood of facing a mimic are consistent with the actual frequency with which mimics try to pass. A strategy for a B is a rule specifying whether to try to pass as a function of his or her effort cost e. A strategy for G is a probability with which to investigate a person emitting A-type signals.

Let β' be the government's belief that it faces a mimic conditional on encountering a person claiming to be an A. Given the government's costs and benefits for failing to detect a mimic, the government prefers to pay the costs of investigation when:

$$\beta' (-pl - d) + (1 - \beta')(-d) > \beta' (-l)$$

$$\beta' > \beta^* \equiv \frac{d}{l(1 - p)}$$

G prefers not to investigate when it believes the chance it faces a mimic is less than this threshold value β^*, and G is indifferent when $\beta' = \beta^*$. Appropriately, the government agent will investigate for a larger range of beliefs β' that it faces a mimic (i.e. when β^* is smaller) when the costs of investigation d are low; the costs of failing to detect, l, are high; and when the ability to detect mimics is high (low p).

Let $\beta'(z)$ be the government's belief that it faces a mimic if all B types with costs of mimicry $e < z$ try to mimic. This is the share of mimics in the set of all people sending A-type signals (Bayes' rule):

$$\beta'(z) = \frac{\beta F(z)}{\beta F(z) + 1 - \beta}$$

Finally, note that $\hat{e} = pv - (1 - p)c$ is the effort cost such that a B with cost $\hat{e}$ for mimicking would be just willing to do it even if she was sure to be investigated. Of course, it can be that $\hat{e} \leq 0$, in which case no B would want to try to pass if a detection effort were certain. However, if there is a high enough chance that you could pass even if investigated (large enough p) or if the benefits of passing are very large relative to the status quo and the cost of getting caught as a mimic, then there can be types of B who would want to try to pass even if the government is investigating everyone.

For example, consider a potential victim of a genocide in progress. If the person believes that he is certain to be killed if he tries to pass and fails, but is *nearly* certain to be killed even if he does not try to pass, then trying to pass even if there is only a small chance it will work may be worthwhile even if he knows that detection will be attempted.[2] Formally, this is a case where c is very close to zero, or equivalently, v is very large.[3]

We are now in a position to describe the game's equilibrium, which is unique for any given set of parameters. Depending on the parameters, patterns of action in equilibrium fall into one of three cases, distinguished by the behaviour of the government. In case (1) below, the government does not bother to investigate anyone, and all types of B with $e < v$ mimic and get away with it. In case (3), by contrast, the government investigates everyone, and only those Bs with effort costs for mimicry less than $\hat{e}$ try to pass. In case (2), the government investigates some but not others.

Proposition 1. The following strategies form a perfect Bayesian equilibrium (PBE) of the mimicry game for three mutually exclusive cases, which depend on the relationship of β^* to two cutpoint values, $\beta'(\hat{e})$ and $\beta'(v)$.[4]

---

[2] For a similar logic applied to free-riding in the context of civil war, see Kalyvas and Kocher (2007).

[3] Recall that we normalized the value of not trying to pass to be zero, so that if the status quo as a B is very bad, then v is large relative to c.

[4] Since $\beta'(z)$ is increasing and $\hat{e} < v$ when $p < 1$, $\beta'(\hat{e}) < \beta'(v)$. In this description of the equilibrium, I will ignore non-generic boundary cases such as $\beta^* = \beta'(v)$.

Case 1 (no one checked): If $\beta^* > \beta'(v)$ then G never investigates, and all types $e < v$ become mimics.

Case 2 (some checked): If $\beta'(v) > \beta^* > \beta'(\hat{e})$ then G investigates any person showing signs of being an A with probability:

$$g^* = \frac{v - e^*}{(1 - p)(v + c)}$$

where e^* is implicitly defined by

$$F(e^*) = \frac{1 - \beta}{\beta} \frac{\beta^*}{1 - \beta^*}$$

B types with $e < e^*$ become mimics and the rest do not.

Case 3 (all checked): If $\beta^* < \beta'(\hat{e})$ then G investigates all people showing signs of being an A, and only Bs with $e < \hat{e}$ try to mimic.

In the first case, either the government finds investigation too costly relative to its expected benefit to make it worthwhile, or there are too few potential mimics in the population to make it worthwhile, or both. The threshold value $\beta'(v)$ is the share of mimics in the population of 'As' if all Bs who would want to mimic if they could get away with it in fact mimic. If this share is small enough relative to the government's threshold belief for being willing to investigate, then G just doesn't bother with mimicry. Of course, within this case, the probability that a B succeeds as a mimic is 1. The number of mimics is increasing in the share of Bs in the population and in the value of passing (v). It is decreasing in the average cost for appearing as an A.[5]

As argued above, for many empirical examples it can make sense to speak of 'the stakes' for mimics and targets. That is, the greater the mimic's benefit for passing, the greater the target's cost for failing to detect; this will be the case, for example, when there is essentially a distributional conflict between mimics and targets. In the model, this means that l is positively related to v. When this is so, increasing the stakes means that more mimicry is attempted while we remain in the bounds of Case 1 (no one checked), but Case 1 becomes less likely to obtain because increasing l means that β^* gets smaller—G is more inclined to investigate when the stakes are larger, of course. And as we will see, increasing the stakes within Case 2 (some checked) has a very different impact on the rate of attempted mimicry.

---

[5] Say that increasing the costs of appearing as an A means a rightward shift in F.

In Case 2, government agents investigate some but not all people claiming to be As.[6] Bs whose costs of mimicry are less than a threshold level $e^* < v$ give it a try. In this case the equilibrium probability that an A is in fact a B is completely determined by the ratio of a government agent's costs of investigating to the expected benefits of investigating a mimic. If there are 'too many' Bs pretending to be As, investigation rates increase. If there are too few, investigation decreases. The share of Bs in the population of apparent As is 'just right' (at β^*) when government agents get the same expected payoff from investigating as they get from not investigating. Note well that this share is completely determined by G's preferences, not B's.

The share of 'A's who are mimics (β^*) is not the same thing as the share of Bs who try mimicry ($F(e^*)$). But it turns out that they act the same way in this case. Since

$$F(e^*) = \frac{1 - \beta}{\beta} \frac{\beta^*}{1 - \beta^*} = \frac{1 - \beta}{\beta} \frac{d}{l(1 - p) - d},$$

the share of Bs who mimic is increasing in G's costs of detection, and decreasing in the danger to G of missing mimics and the probability of getting caught. This all seems intuitive. A non-obvious implication, however, is that within Case 2, *increasing the stakes lowers the rate of mimicry*, since increasing l lowers the share of Bs who mimic (and the share of 'A's who are actually Bs), while increasing the importance of passing for Bs, v, has no effect at all.

At the same time, because increasing the stakes makes G's threshold belief for investigation, β^*, smaller, it becomes more likely that Case 3 (all checked) obtains rather than Case 2 (some checked). As discussed below, in Case 3 G investigates everyone and more Bs try mimicry as the stakes increase. So the effect of increasing the stakes on the rate of mimicry is complicated. It is increasing when the government does not care enough to investigate potential mimics, decreasing when the government wants to investigate some but not all, and then increasing again when the government cares enough to want to check all As.

Besides the stakes, the share of Bs who try mimicry is also determined by the size of the population of Bs. In Case 2, the larger the group of potential mimics, the smaller the share of them who try mimicry. So if we begin with an extremely small group (β close to zero), G initially doesn't care about a few

---

[6] In a slightly more realistic formulation, mimics could differ randomly in the degree of suspicion they raise when trying to pass, or government agents could differ in their degree of perceptiveness. Thus the agents would not need to adopt a consciously mixed strategy, but it would be effectively random from the perspective of Bs.

mimics (Case 1). But eventually, for a large enough group of Bs, investigations begin and absolutely fewer Bs will mimic as the size of the group grows.

In Case 2, the government's rate of investigation of 'A's (g^*) has the following comparative statics:

- increasing in the stakes
- increasing in the probability that a mimic passes if investigated
- decreasing in a mimic's costs for getting caught
- decreasing in the costs of investigation, and
- decreasing in the share of the population that are Bs.

Because not every 'A' is investigated in Case 2, the probability of successful passing is greater than p, the chance of passing when investigated. A mimic's unconditional probability of passing in this case is $1 - g^* + g^*p$ which equals $(e^* + c)/(v + c)$, and has the following comparative statics. Successful passing is more likely when:

- the costs of investigation are higher
- the probability of passing if investigated (p) is higher
- the costs of getting caught, relative to the benefit of passing, are *higher* (this is because fewer attempt mimicry)
- and the stakes are lower.[7]

Finally, we come to Case 3 (all checked). Here, even if the government is going to investigate every single person claiming to be an A, there are enough Bs who want to try to pass to make comprehensive investigation worthwhile for the government. This is more likely if the expected savings for the government by investigating are large relative to the costs of detection—that is, $l(1 - p)$ is large relative to d—while the expected benefits of attempting to pass are large for a mimic ($pv - (1 - p)c$)). For this situation to obtain, investigation must not be too accurate—if it were then not enough Bs would want to try mimicry for comprehensive investigation to be worthwhile—nor too inaccurate—if it were then it would not be efficient for the government to investigate everyone.

So, provided that investigation is of at best medium accuracy, raising the stakes high enough pushes the interaction towards Case 3 by increasing the share and number of Bs attempting to pass even if they are sure they will be investigated.

---

[7] e^* is decreasing in v when l is positively related to v.

Figure 2.1 and Figure 2.2 summarize the most important of the comparative statics results described above. Figure 2.1 shows how several measures of the extent of mimicry vary with the stakes at issue. Figure 2.2 shows how the equilibrium shares of mimics from the Bs, among the As, and in the total population vary with the size of the B group. The three cases are visible in each figure by the breakpoints where the slopes of the lines change abruptly. For example, in Figure 2.2, when β's are fewer than about 20 per cent of the population (given the parameters used to generate these figures), all Bs mimic—this is Case 1 (no one checked). The intermediate range is Case 2 (some checked), and the curves after stakes of 1.5 in Figure 2.1 and β of about .66 in Figure 2.2 are Case 3 (all checked).

Figure 2.1 illustrates how mimicry increases with the stakes until the government is motivated to start investigating, at which point increasing the stakes actually reduces mimicry rates, up to the point where the government wants to investigate all claiming to possess the desirable characteristic

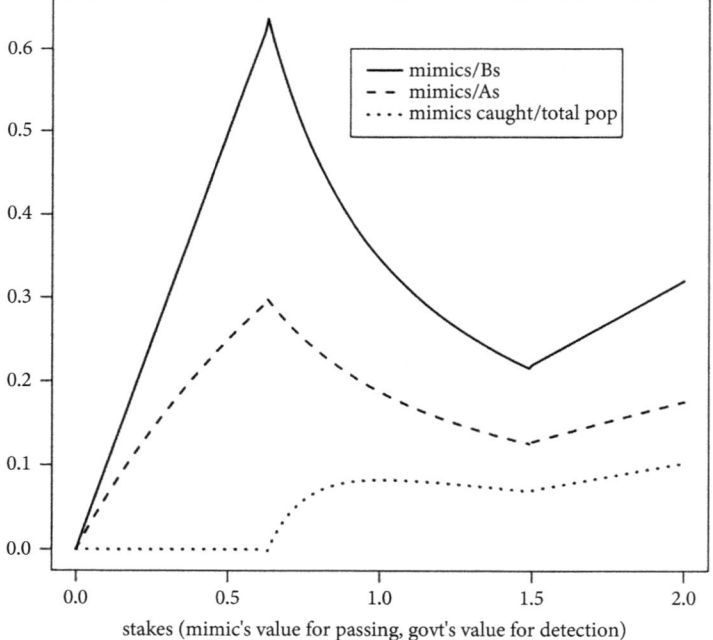

Figure 2.1 Share of Bs who mimic as function of stakes
Source: created by author

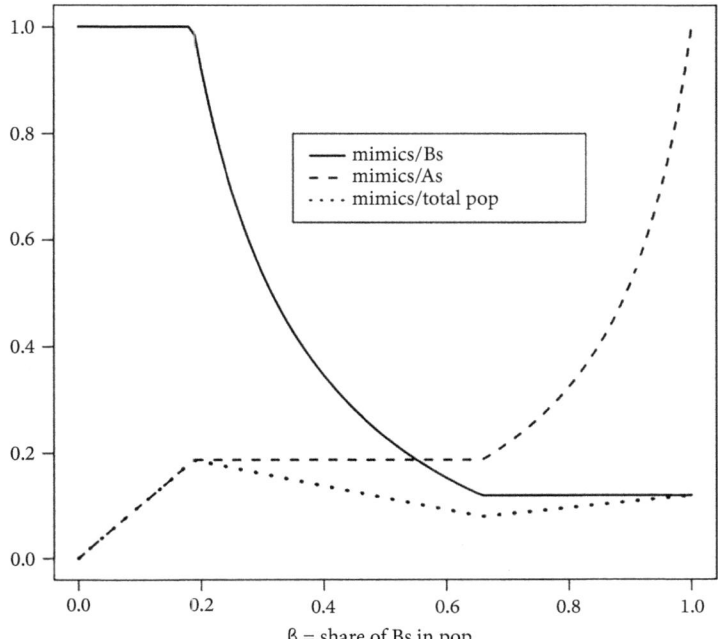

Figure 2.2 How mimicry varies with population share of possible mimics
Source: created by author

A. Then greater stakes increase mimicry again as more *B*s give it a try even though they are sure to face investigation.

An interesting if unfortunate implication of this result is that it suggests that it may be hard to identify any robust empirical patterns concerning the size of the stakes and the scale of mimicry. Apart from the observation that mimicry will be common and unpoliced if there are few potential mimics and they can do little damage from the target's perspective, mimicry efforts may be rampant or quite uncommon when the stakes at issue are moderate or high. Much depends on how accurate investigations are.

Figure 2.2 suggests a simpler pattern: the share of *B*s who choose to try mimicry weakly decreases with the size of the *B* population. For some empirical settings, this could be a testable prediction.[8] By contrast, the share of mimics among the '*A*'s is increasing at first, then stable, and then increasing again as the size of the *B* group grows.

---

[8] For example, did Nazi-occupied European countries, or districts within a country, with more Jews have lower rates of mimicry? There would of course be an 'other things equal' problem here in that stakes might also vary with population shares.

Some Evolutionary Dynamics

In the equilibrium of the generic social mimicry game analysed above, government officials do not want to investigate either more or less given the true number of mimics, and Bs do not want to mimic at a higher or lower rate given the true probability of being investigated and passing. I gestured at an argument that decentralized decision-making would actually yield such an equilibrium: if there are too many mimics, government officials would increase detection efforts, reducing mimicry; if 'too much' detection effort, fewer mimics, and thus a switch to less detection.

But is that right? If we make reasonable assumptions about what G and Bs can observe, and about what information they use in adjusting their strategies, would the implied dynamical system actually converge to the equilibrium identified above?

Suppose that the action plays out over a series of periods $t = 0, 1, 2, \ldots$, which could be days, weeks, months, etc. In $t = 0$, the government agent G randomly chooses $g_0 \in (0, 1)$, a share of people presenting themselves as As to investigate, while Bs with effort cost less than a randomly determined level $\hat{e}_0$ attempt to pass. G then learns the percentage of people she checked whom she determined to be Bs trying to pass. If she knows (or can estimate relatively well) the probability that she can correctly spot a true B who is trying to mimic ($1 - p$ in the model), then she can 'back out' an estimate of the percentage of people who presented as As who actually are Bs simply by dividing the number of Bs she identified by $1 - p$. This is β' in the model above, so call the first-period estimate β'_0 in this dynamic version.

The government agent has a notion of whether 'too many' Bs are trying to pass. In the model as described above, if $\beta' > \beta^*$, then the agent finds it worthwhile to investigate more As, because the expected costs they impose, or the expected penalties for the agent from letting too many Bs through, are greater than the costs of investigating another A (those costs are d per individual). So it seems reasonable to suppose that in the next period, G will increase the frequency of investigations if $\beta'_0 > \beta^*$, and reduce the frequency if $\beta'_0 < \beta^*$. A simple adjustment rule for periods $t > 0$ is:

$$g_t = g_{t-1} + \varepsilon \left(\beta'_{t-1} - \beta' \right), \tag{2}$$

where $\varepsilon > 0$ is a parameter that reflects the amount by which G changes investigation frequencies. This proposal makes the plausible assumption that G makes smaller changes when the rate of B's trying to pass is closer to the rate at which G is just indifferent between doing and not doing another investigation.[9]

What about the Bs? For many contexts it would be reasonable to suppose that they can get information about the rate at which Bs who try to pass are investigated and the rate at which they succeed. This would be the case, for example, if Bs are a relatively small ethnic or religious minority faced with persecution, so that word spreads in the community about who tried to pass, who got away with it, and who got caught. Teenagers in a large senior school get some information about how many peers have fake IDs and how often they work. The Internal Revenue Service in the US (IRS) advertises its audit rate and it is probably reasonable to assume that the probability of 'passing' in that case is quite low. Villagers in Columbia could update about the military's screening policies—and what worked or did not work in terms of falsifying ID cards to reduce the amount of time spent renewing them—fairly rapidly by just talking with other villagers.[10]

To the extent this is the case, a simple and natural adjustment rule for the Bs is that they 'best respond' to the detection frequency of the previous period. A type with cost for mimicking e thus estimates an expected payoff for trying to pass of $v(1 - g_{t-1} + g_{t-1}p) - g_{t-1}(1 - p)c - e$, which implies that in period $t > 0$, types e less than the threshold value:

$$\hat{e}(g_{t-1}) = v(1 - g_{t-1} + g_{t-1}p) - g_{t-1}(1 - p)c \qquad (3)$$

will mimic.[11]

To close the model, note that β'_t is given by (1) using $\hat{e}(g_{t-1})$ for z. This system is easily simulated or considered analytically. If the government agent's adjustment rate is not so large that she (pathologically) bounces between investigating everyone and investigating no one, then the system converges rapidly to the equilibrium levels of g^* and e^* found in section 3. Figure 2.3a and Figure 2.3b illustrate two cases, where in the second case Gs adjust the frequency of investigation in large increments, 'overshooting' and producing some oscillation. In this example, in both cases G starts out investigating at a rate that deters all Bs from attempting to mimic, which leads G to investigate

---

[9] Since g_t is a proportion, I need to add that we should take the median of 0, 1, and the value given by (2).

[10] Personal communication with Diego Gambetta, relaying information from Juan Masullo, based on the latter's research on civilian responses in the Middle Magdalena region north of Bogotá.

[11] Here we need to add the technical stipulation that if $\hat{e}(g_{t-1})$ is greater than the upper bound on B's possible costs for trying to pass (if there is one), we set $\hat{e}$ in t to that upper bound, and likewise to zero if the value in (3) is less than zero.

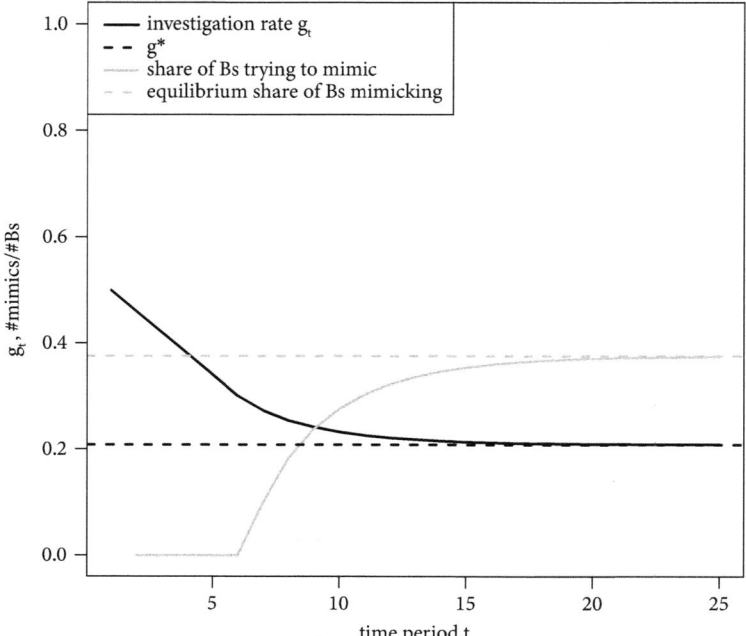

Figure 2.3a 'Slow' Adjustment

Source: created by author

less because there is no one to catch, so *A*s are just being inconvenienced. Soon enough it begins to be worthwhile for the lowest cost *B*'s to start trying to pass.[12]

Possibly the strongest assumption made above is that the government agent can reliably estimate her own probability of correctly detecting a mimic. In most contexts, a successful mimic leaves no trace. If the government, or 'dupes', are unable to observe the size of the population of successful mimics, then it is not clear how they can estimate how good they are at spotting them. If they overestimate their ability to detect mimics—which seems psychologically plausible to me on the grounds that we have a tendency not to think about 'dogs that didn't bark'—then *B*s will be able to take advantage of this and will mimic at higher rates than e^*, while *G* investigates too little. This is most likely if the *B*s are a cohesive group with strong social networks, which would make for effective information flows about the probability of being investigated and the ability to pass. In civil war contexts, villagers

---

[12] Parameter values are $v = 1$, $c = 5$, $p = .5$, $\beta = .4$, $\beta^* = .2$, and e is uniformly distributed on [0, 1]. In Figure 2.3a the adjustment rate ε is .2 while in Figure 2.3b it is 1.5.

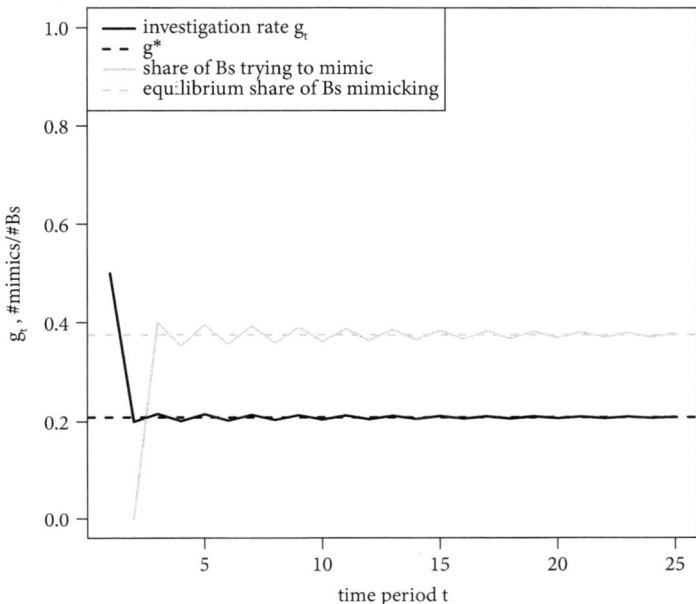

Figure 2.3b 'Fast' Adjustment

Source: created by author

would generally be expected to have good information about passing rates in interactions with military and rebel forces.

On the other hand, if the Bs are not a cohesive group and if the penalties for being caught trying to mimic are very large, then the reverse might well obtain. That is, very few attempt to pass and little is known about success rates, which means that Bs have little to go on and so may not try to mimic despite the fact that if they did, they would be quite likely to pass. For example, there may be a great many opportunities to successfully and profitably mimic that arise in everyday life, but 'too few' take them due to underestimates of the ability to get away with it.

Institutional Commitments to Investigate Potential Mimics

In the game analysed above, the target chooses whether to investigate for mimicry on a case-by-case basis. It does not have the option of committing ex ante to investigate all people claiming to be As. In some empirical settings—particularly those where the targets are government agents—it is natural to suppose that the government can make institutional commitments on the frequency of investigation. For instance, the US Congress can fund the Transportation Security Administration and pass laws committing the TSA to screen all airline passengers. The IRS can commit, for the most part, to implement a particular rate of auditing tax returns. Customs systems are institutional commitments to investigate the citizenship claims of people entering a country. In other settings, however, institutional commitment is more difficult. The government cannot commit sales staff in liquor stores to check all customers' IDs, for instance. Sometimes in situations of official efforts to prevent mimicry, the government faces a principal–agent problem in figuring out how to motivate and monitor its agents' efforts to detect mimics.

Consider then a modified version of the signalling game in which the target moves first, choosing whether to commit to investigate all people claiming to be an A, or to investigate on a case-by-case basis as above.[13] Suppose for starters that the institutional commitment to investigate all is itself costless (although the larger number of investigations pursued will be costly).

Proposition 2. If G can commit in advance to investigate all persons claiming to be As, then no commitment is made if the conditions for Case 1 (no one checked) or Case 3 (all checked) in Proposition 1 obtain, while commitment is made if the conditions for Case 2 (some checked) obtain. Equilibrium strategies are then the same for Cases 1 and 3, while in Case 2 the government investigates everyone, and Bs attempt to mimic only if their costs of doing so are $e < \hat{e}$ (as in Case 3).

Proof. Without commitment, in Case 2 G gets an expected payoff per encounter $-pl\,\beta^* - d$. By committing ex ante, G induces all types of B with $e > \hat{e}$ to not mimic, yielding an expected payoff per encounter of $-pl\beta'(\hat{e}) - d$. Since in Case 2 $\beta^* > \beta'(\hat{e})$, this means that commitment is always better for G in this case.

In Case 1, G's expected payoff per encounter is $-l\beta'(v)$, and by committing it could get $-pl\beta'(\hat{e}) - d$. Rearranging, commitment is worse for G than no commitment in this case when $\beta'(v) - p\beta'(\hat{e}) < d/l$. This is certainly true when $p = 0$, since then $\beta'(v) < d/l = \beta^*$ is a condition for Case 1 when $p = 0$. Since $\beta'(v) - p\beta'(\hat{e})$ is decreasing in p, commitment is worse than no commitment for G for any p when the condition for Case 1 holds.

---

[13] A slightly more general version would allow the government to commit to any given probability of investigation g, but I will consider the full investigation commitments versus no commitment for the sake of clarity.

For Case 3 parameters, commitment by G is superfluous since investigation is comprehensive in this case anyway.

So if it were costless to set up, the government would always want to construct ex ante institutional commitments to investigate all 'A's if it has an incentive to investigate some. The reason is that there is a gain in terms of deterrence of mimicry. G investigates more often than without commitment, but this is more than compensated by the reduction in mimics who get through.

Why then is commitment necessary? Because after setting up the institutions to force investigation of all presenting as As, the rate of mimicry drops to a point where G would like to stop investigating so much—this is why commitment is needed. If the TSA and other airport precautions are highly effective, then there is a lot of wasted cost and effort, ex post.

Proposition 2 assumed for clarity that there are no costs to setting up an institutional system of commitment to investigate apparent As. If we introduce costs of building a monitoring system, then whether G wants to do so in Case 2 will depend on a comparison of the costs of institutional commitment versus the gains from deterring more mimics. Higher stakes, or an increase in the number of potential mimics will tend to favour moves towards institutional commitment to investigate.

Illustrations

Fake IDs

The costs of procuring a fake ID are presumably small compared to the value of having one if it is successful (v), but the cost of getting caught (c), a minor felony, would be large for many or most kids. Then in a Case 2 equilibrium (some checking), young-looking customers would be checked at a rate of about $v/(v + c)$, which one would think would be less than one half.[14]

The costs of checking ID for a liquor store shop worker are small but not nothing—possible loss of sale, possible irritation of of-age buyers, discomfort at implying dishonesty—while, at least for the store owner, the expected cost of selling to minors might be quite large (risk of losing licence). If so, then in a Case 2 equilibrium we would expect a low rate of fake ID use among

---

[14] Who are the Bs and who are the As here? For a good part of the population, visual inspection can determine with high confidence that the person is over 21, or under, say, 12. The As would be everyone else who is over 20, and the Bs those who are less than 21. A model tailored to this specific problem would incorporate a person-specific signal of age.

late teens. Further, if the costs of selling to underage buyers are large enough for the store owners that checking IDs is optimal, then they may want to have a policy of 'We Check Everyone' to try to commit clerks ex ante, as discussed above. Psychologically, this could also lower the costs of checking for the clerk, by mitigating the specific implication of dishonesty.

Taxpayers

Falsifying a tax return is practically costless in and of itself. The gain to getting away with it (v) is the tax rate times the amount of income underreported. The cost of getting caught is some sort of large financial penalty and maybe worse. Many also probably factor in reputational damage as a cost of being found guilty of cheating on tax returns.

This would again imply a low rate of equilibrium auditing by the IRS (that is, $v/(v + c)$ would be small). But the true rate of auditing is in fact *extremely* small—on the order of one in 130 returns in the US—so most people would have to view the costs of getting caught as 130 times the value of cheating to rationalize this audit rate. Further, I believe I have read that the net financial yield on additional IRS audits in terms of penalties and delinquent taxes is very high, which would not be the case if the IRS was simply trying to maximize its (legal) take. Most likely their principal, Congress, does not want too many audits because these are unpopular with constituents.

We have good evidence that this should be a Case 2 equilibrium, since surely the equilibrium cannot involve auditing of everyone or no one. So without institutional commitment the rate of cheating on tax returns (mimicry) would in equilibrium be determined by the ratio of the IRS's political and financial costs for an audit to the expected political and financial returns from an audit. It is not obvious to me if this would be a small number (less than 20 per cent?), or something larger. However, this is also a case where it seems plausible that the IRS can commit in advance to audit a given percentage of all tax returns.

IRA Operatives

In this example (and possibly the fake ID example above), a more natural formulation of the model would have Catholics in Northern Ireland varying across individuals in how much they would value being an IRA operative if they could get away with it, and pay no social or other costs for becoming one.

That is, the individual variation would be on the benefit of passing v rather than on the cost of becoming an operative e. While some of the expressions in the formal analysis change slightly, there are no substantive differences in results. Types of B with a value for being an IRA operative greater than a threshold value v^* opt for the organization, while the rest do not.[15]

IRA operatives would probably be hard to detect by interrogation of random individuals, and the financial and political costs of individual interrogations are presumably at least moderate relative to the expected damage foregone by catching an operative (high p, moderate d relative to l). This would suggest Case 1: not enough Catholics who want to pay the cost e to be an operative to make individual interrogations of random Catholics worthwhile for the government. This could differ in neighbourhoods or bounded social sets where many operatives are suspected (higher β).

Potential Victims of Genocide

In the midst of an active genocide, the options for a B may reduce to: (1) wait for neighbours and government agents to identify and kill you, with high probability, or (2) pre-emptively seek to escape the country by passing as an A, though facing certain death if you fail to pass. Given these options, it is surprising that, as far as we know, more European Jews did not pre-emptively seek to escape the Nazis by passing, and more Tutsis did not try to pass as Hutus in Rwanda. It should be stressed that this is a stylized fact at best. I do not think we know much about actual rates of attempted passing and their success in either case, especially for the latter. Also, since we know what ultimately happened, it is easy for us to overestimate how confident Jews and Tutsis should have been at the time that not mimicking would lead to death. Nonetheless, as Laitin (in this volume) argues, a fair bit of evidence suggests the rates of successful and attempted mimicry cannot have been large, which seems surprising.

What does the model imply for plausible parameter values in this example? The stakes are clearly very large, which would put us in Case 3 if B's chances of passing are not negligible. In Case 3 all are investigated and most Bs try to pass. However, if the probability of passing, p, is quite small, so that other options like hiding, bribery, or seeking protection from friends would be more attractive if you were certain to be investigated for mimicry

---

[15] In Case 2 with this formulation, types of B with $v > v^*$ mimic, where v^* is implicitly defined by $1 - F(v^*) = \frac{1-\beta}{\beta} \frac{\beta^*}{1-\beta^*}$. G investigates with probability $g^* = (v^* - e) / (v^* + c)$.

$(pv − (1 − p)c < 0)$, then we return to Case 2. Here, because the stakes are large for the government relative to the costs of investigating (it wants to kill these people after all), the equilibrium rate of mimicry is predicted to be very small, even if not everyone is thoroughly investigated.

The evidence Laitin presents suggests that, at least for the Rwandan case, the probability of successful passing if investigated was in fact extremely small. The main problem seems to have been that in a rural, village-based society, everybody knows your name, appearance, and history, so there were abundant sources of information for *génocidaires*.[16] In addition, as Laitin discusses, the government had for years invested in institutions that facilitated screening—perhaps in line with the results on institutional commitment above. Exiting the village and trying to pass at multiple roadblocks appears also to have been highly unlikely to work, in large part because the government had for many years required individuals to carry official ID cards that specified tribe.[17] The prediction from the model would be high but not comprehensive investigation rates, and low levels of attempted mimicry.

Passing also appears to have been extremely difficult for the most part in Nazi-occupied Europe. For men and boys, circumcision was nearly impossible to disguise.[18] Women and girls did not face this handicap, and Laitin cites a study by Weitzman that claims that 'thousands' of Jews, 'mostly women', escaped the Holocaust by mimicry. The difficulty of acting independently of a family unit (with males) may help explain why this number is not even higher, but it may still be surprising that it is not larger.

Conclusion

In Michael Spence's (1973) classic signalling model, there are good workers and bad workers, the former having higher productivity than the latter. Good types have a lower effort cost for years in school, and, in order to starkly illustrate the logic of signalling, Spence assumes that education has no effect on productivity. In a separating equilibrium, good types choose a level of education that is high enough that bad types would not want to mimic it

---

[16] This raises a more general point: Social mimicry simply cannot be much of an issue except in cities and mass societies.

[17] Fake IDs were hard to procure on short notice. As Laitin observes, a big puzzle in this case is why more Tutsis had not prepared fake IDs long before in anticipation of such dangers.

[18] From the website 'Virtual Jewish Library': 'Because non-Jews in continental Europe generally were not circumcised, German and collaborationist police commonly checked males apprehended in raids. For boys attempting to hide their Jewish identity, using a public restroom or participating in sports could lead to their discovery. More rarely, they underwent painful procedures to disguise the mark of circumcision or even dressed as girls.' See also Gambetta (2005: 232) on how cultural knowledge made it extremely difficult for Polish Jews to pass as Christians.

even though doing so would get them a higher wage from employers. This outcome is inefficient. The education costs are a pure loss relative to the first-best outcome that would occur if employers could observe ability directly. The information asymmetry drives good types to incur education costs to distinguish themselves from potential mimics.

In this example, then, a whole system of institutions—higher education— is explained in part by a problem arising from the potential for mimicry (or, in economic terms, an information asymmetry in a situation where there is an incentive to misrepresent). What colleges and universities do, in this argument, is not so much to add value in terms of ability but to screen good from worse types in the admissions process.

Now consider the contrast between social mimicry as described by Gambetta and as modelled above, and the classic signalling model. In the former, the focus is on strategic behaviour by the potential mimics and dupes: do they mimic or not? Do they investigate or not? In the latter, the focus is on strategic behaviour by the people the mimics want to mimic, or 'models' in Gambetta's language: how can they credibly distinguish themselves from other types? This is not particular to Spence's example. Whenever there is a benefit that many desire but that politics, ethics, or economic efficiency wants to allocate to some subset, then 'models' want to be able to distinguish themselves from mimics.[19]

In the approach suggested by the standard signalling game, we would study social institutions with an eye to how they are constructed to screen out potential mimics, to be 'incentive compatible', in the language of economics. Social mimicry would then be understood as an instance of, or another way of describing, the problem posed by private information and incentives to misrepresent.

That is, a large number of social, political, and economic interactions have the following aspect. What should be done to produce a good outcome for a group or for an individual should ideally depend on private information held by different people. But because of distributional conflicts, people might misrepresent their private information if they were simply asked. The deep structure of such problems has been mapped and explored by economic theorists under the rubric of 'mechanism design'.[20]

---

[19] It is worthwhile spelling out exactly how the classic Spence model differs from the mimicry model analysed here. In the Spence model, the employer (or G in the model here) is completely incapable of figuring out if the job applicant is a high or low productivity type except by actually putting them to work; investigation is useless ($p = 1$). However, in Spence's case, job applicants can verifiably report their effort level e, and the marginal cost of effort is higher for low productivity types than for high productivity types. Thus this is a class of situations where the 'models'(as Gambetta terms them) are able and have an incentive to take actions that verifiably distinguish themselves from potential mimics.

[20] For an introduction, see the Nobel lectures of Maskin (2008) and Myerson (2008).

A mechanism can be thought of as a formal or informal institution that takes as inputs some form of 'announcements' of an individual's private information (often called 'types') and maps these into social decisions or outcomes. Tax systems, voting systems, legislatures, juries, markets, auctions, pricing policies by businesses, many websites, school systems, political constitutions, all manner of contracts, police departments, counter-insurgency methods—all these and much more are mechanisms. From the perspective suggested by the mechanism design approach, institutions elicit signals and screen among different types, producing various collective decisions as results. Institutions may be informal, as with customs concerning how buyer–seller bargaining proceeds, or highly formal, as with how a political constitution structures bargaining among branches of government.

Societies that have managed to get to the 'mass' level with cities and lots of impersonal exchange have done so because they have developed institutions that make social mimicry appear to be a marginal problem at best, in that almost all the time we can trust the representations of others we deal with on a daily basis. But this is precisely because we have developed formal and informal institutions that deter mimicry in anonymous and 'one shot' interactions. Seen in this light, solving the problems posed by potential mimicry is a central challenge for low-income countries trying to develop economic and political institutions that will support complex economic exchange among larger-than-traditional groups of people and organizations, and political institutions wherein voters can trust politicians who pretend to be 'good types' to actually be good types. High-income countries have developed institutions and cultures that resolve this problem to a greater degree, although new technologies like the Internet pose new threats by creating new ways to mimic, as Hegghammer (in this volume) discusses for the case of Jihadi Internet forums.

References

Gambetta, Diego. 2005. 'Deceptive Mimicry in Humans'. In *Perspectives on Imitation: From Neuroscience to Social Science*. Volume 2: *Imitation, Human Development and Culture*, edited by Susan L. Hurley and Nick Chater, 221–242. Cambridge, MA: MIT Press.

Kalyvas, Stathis N. and Matthew Adam Kocher. 2007. 'How "Free" Is Free Riding in Civil Wars?: Violence, Insurgency, and the Collective Action Problem'. *World Politics* 59 (2): 177–216.

Maskin, Eric. 2008. 'Mechanism Design: How to Implement Social Goals'. *American Economic Review* 98 (3): 567–576.

Myerson, Roger. 2008. 'Perspectives on Mechanism Design'. *American Economic Review* 98 (3): 586–603.

Spence, Michael A. 1973. 'Job Market Signaling'. *Quarterly Journal of Economics* 87 (3): 355–374.

3

Can You Trust Anyone on Jihadi Internet Forums?

Thomas Hegghammer

> Ask yourself, why does this person trust you? In a world where liter-
> ally anyone can be a spy, why does this person trust you? Why are
> they claiming to be a Mujahid, and telling this to a person who they
> met over a computer or at the masjid? Why are they telling you they
> want to conduct Jihadist operations or make hijrah? You cannot know
> if anyone is sincere, and this is the sad reality. Think about why they
> would trust you of all people, and not someone else. The answer is
> because they are seeking to arrest you.
>
> **From '10 Methods To Detect And Foil The Plots Of Spies',**
> *Ansar al-Mujahideen Forum*, **19 December 2010**

The Internet is often believed to empower international terrorists by enabling
them to communicate rapidly and cheaply across long distances.[1] But
cyberspace is also a fertile arena for trickery and deception, because it is easier to hide your true identity online than offline. This chapter looks at how
the threat of deceptive identity mimicry affected radical Islamist discussion
forums in the 2000s and early 2010s. Rather than serving as safe havens for
plotting and recruitment, jihadi forums were dens of distrust and paranoia.[2]
Users were extremely suspicious of one another and rarely engaged in sensitive transactions. Activists tried to mitigate the problem with vetting agencies,
reputation systems, complex document formats and sorting knowledge, but

---

[1] I am grateful to Diego Gambetta for detailed comments on early drafts of the manuscript. Dorothy
Denning, Lynn Eden, James Fearon, Tim Junio, Christopher Sullivan, and Aaron Weisburd also con-
tributed constructive remarks. I thank all who participated in the seminars discussing the paper at the
Stiftung Wissenschaft und Politik in Berlin, Stanford University, the Naval Postgraduate School, UC
Berkeley, Lawrence Livermore National Laboratories, and Yale University.
[2] 'Jihadi'is an adjective derived from 'jihadism' which I equate with 'violent Sunni Islamism'. I define
Islamism as 'activism justified with primary reference to Islam.'

Thomas Hegghammer, *Can You Trust Anyone on Jihadi Internet Forums?* In: *Fight, Flight, Mimic.*
Edited by: Diego Gambetta and Thomas Hegghammer, Oxford University Press. © Thomas Hegghammer (2024).
DOI: 10.1093/9780191802454.003.0003

trust remained fragile. These insights have direct implications for counterterrorism policy and indirect relevance for opposition activism in authoritarian states.

The inquiry speaks to a long-running scholarly debate over the Internet's usefulness for political activists. Technological optimists emphasize the Web's advantages, arguing that it empowers non-state actors and facilitates transnational activism (Arquilla and Ronfeldt 2001; Castells 2009; Tarrow 2005). A substantial literature expounds on the Internet's many presumed benefits for international terrorists in particular (Nacos 2002; Seib and Janbek 2011; Weimann 2006). Technological pessimists, by contrast, highlight the risks and problems terrorists face online (Amble 2012; Conway 2006; Diani 2000; Kenney 2010; Torres Soriano 2012; Zanini and Edwards 2001). This chapter leans to the pessimist side by highlighting the trust problem as a major hurdle in rebel exploitation of the Internet. Interpersonal trust, or 'the willingness to accept vulnerability or risk based on expectations regarding another person's behavior' (Borum 2010), is a vital prerequisite for human cooperation (Misztal 1996). In cyberspace, however, trust is fragile because the absence of non-verbal cues makes it easier to lie. Hunted activists must be especially cautious, lest they fall victim to deceptive mimicry from a government agent provocateur or other types of 'faux activists' (Marx 1974; 2013). The threat of digital surveillance adds to their woes.

The chapter is an empirical probe into the effects of the trust problem on the social interactions between militant Islamists online. From the point of view of the framework of this volume, it looks at countermimicry, i.e. the strategies and techniques of online jihadists to defend against the offensive mimicry efforts of police infiltrators.[3] I examine pro-al-Qaida discussion forums from the late 2000s and early 2010s, asking three main questions: how much trust was there between users? What types of signs instilled trust and distrust respectively? Which strategies did users follow to guard against deception? I also hope to shed light, in the process, on broader questions in trust research, including: can trust arise at all among high-risk activists in virtual settings? What happens to the 'semiotics of signs' (Bacharach and Gambetta 2003) when all signs are virtual? What makes a 'costly sign' in cyberspace? We lack answers to these questions because, with the exception of a few works on cybercrime (Lusthaus 2012), the online trust literature has been mostly concerned with low-risk activism.

I chose jihadi forums because it is an arena where the trust problem should be particularly acute and its effects therefore more easily observable. Jihadi

---

[3] For insights into the jihadists' own offensive mimicry efforts, see Milton (2022).

forums are also important in their own right, because they have been central to the al-Qaida movement in the post-9/11 era. For over a decade, this was 'where the global jihad is headquartered online' (Zelin 2013, 6). I use open-source messaging data which I process qualitatively with content analysis. The data have many limitations, but they allow at least for some tentative insights into the internal communications of an otherwise highly clandestine community.

The chapter has eight parts. First I describe what a forum is; then I review the literature on online trust before presenting my sources and methods. In the fourth part I assess the level of government repression on the forums. The fifth part assesses the overall level of trust, the sixth looks at which signs make and break trust, and the seventh presents some common mitigation strategies. The final section concludes.

Jihadi Discussion Forums

Radical Islamists established a presence on the Internet in the mid-1990s, and since the early 2000s they have been at the forefront of rebel Internet use, exploiting it more extensively than most other types of militants (Brachman 2006; Bunt 2003; Bunt 2009; Conway 2006; Drennan and Black 2007; Hoffman 2006; Awan, Hoskins, and O'Loughlin 2011; Lappin 2011; Ranstorp 2004; Ranstorp 2007; Rogan 2006a; Rogan 2006b; Seib 2008; Seib and Janbek 2011; Thomas 2003). Their presence can be seen all over the World Wide Web, from static websites via blogs and forums to social media (Zelin 2013). In addition, they communicate by email, messaging services such as Whatsapp, and many other platforms. The world of online jihadism is complex and in constant flux; websites come and go, new technologies replace old ones, and the actor landscape changes (Kimmage 2008; Lia 2006; Weisburd 2007).

Discussion forums were adopted as a communication platform by jihadists around 2000 and remained their most important online meeting place from the early 2000s to the early 2010s. At that point they were superseded by social media such as Facebook, Twitter, and later, Telegram as the jihadi distribution platform of choice. As of 2023, some forums still exist, but they are less central to online jihadism than services such as Telegram. This chapter is primarily concerned with what we may call the 'golden age' of the jihadi forums, that is, from the early 2000s to the early 2010s.

A discussion forum is basically a digital noticeboard where users can post anything from one-word messages to elaborate propaganda products (ICT/JWMG Team 2012; Musawi 2010). Access is generally open: all forums allow users to register with a username and password. Some forums are also accessible to non-registered visitors, while others forums require users to log on in order to access parts (and occasionally all) of the forum. This barrier is sometimes described by observers as password-protection, which is partly misleading because anyone can register and choose their own password.[4] The main purpose of the registration is to allow reidentification and thereby the existence of reputation systems, as we shall see later on in this chapter. Registration has occasionally also been used to maintain a tier-based access system whereby parts of the forum are openly accessible, while other parts are only available to selected user profiles.

Forums are made up of individual messages or 'posts'. Any user can write a new post, and any user can add follow-up posts to existing ones, thereby creating 'threads' (or 'conversations'). Threads have headlines and are listed chronologically, the most recent on top, in a 'section'. A jihadi forum typically has five to ten different sections on different topics such as 'news', 'statements', 'Sharia matters', 'tactical matters', and the like. A section page (see Figure 3.1) is typically divided into three areas: the top part is filled with graphic banners advertising films and statements by major groups. The second part lists 'sticky threads', i.e. the ten to twenty threads deemed by forum administrators to merit extra attention.[5] The third part of the page lists the regular threads.

Figure 3.1 Screenshot of *al-Shumukh* forum, 3 February 2012 ('sticky posts' in upper section)

---

[4] There have been instances of forums ceasing to accept new registrations, in which case access was limited to existing account holders. However, existing account holders still numbered in the thousands and were not screened.

[5] They stick to the upper part of the page instead of being pushed down by new messages.

In the heyday of the jihadi forums, there were between five and fifteen jihadi forums in operation at any one time, although traffic tended to cluster in one or two market leaders.[6] The forum ecosystem was in constant flux; aside from URLs changing up to several times a year, a forum rarely existed for more than a few years (sometimes only months) before disappearing without explanation.[7] We lack reliable traffic data, but messaging activity suggests that at their height, the top forums were frequented by at least several hundred individuals every day. Activity appears to have increased markedly in the second half of the 2000s: the aggregate annual number of threads on the al-Falluja forum, the market leader for most of this period, increased from 3000 in 2006, 18,000 in 2007, and 19,000 in 2008, to 56,000 in 2009.[8] Subsequent activity appears to have stagnated at the 2009 level. In the spring of 2012, Zelin (2013) observed around 150 threads per day (corresponding to around 55,000 per year) on the al-Shumukh forum (the new market leader). According to Zelin's count, al-Shumukh saw between 1500 and 2000 posts per day in early 2012, while the number two forum (al-Fida') had between 700 and 900 posts per day.

We cannot know exactly who the users are, but at least four ideal type categories appear to be represented. First and most numerous are the 'consumers', the young unaffiliated Islamists who access forums primarily to read, exchange, and comment on news and propaganda products. Second most numerous are the 'propagandists', whose main preoccupation is the distribution of ideological documents. Third and fewer in numbers are the 'operatives', that is, the members of active groups who seek to obtain operationally useful information and conceivably to recruit. Fourth is an unknown number of 'impostors', consisting of the intelligence operatives, journalists, and academics who listen in on the above while staying silent or mimicking jihadists.[9] Virtually all forum users participate under pseudonyms, with the exception of a few government actors (notably the US State Department)

[6] For names and URLs of jihadi forums from 2009 onwards, see the article series titled 'Top Ten Jihadi Forums' on Aaron Weisburd's *Internet Haganah* website; for example at http://internet-haganah.com/harchives/007253.html (accessed 16 February 2013).

[7] When the URL of a forum breaks, users wait until the new address is announced on other forums or websites. The re-emergence can take hours, days, or even weeks, and meanwhile, users cannot know whether the forum is simply moving or gone for good. The cause of a forum's disappearance is almost never publicly known, but candidate explanations include hacking by governments and hacker groups, arrest of the site administrators, and eviction by the Internet service provider.

[8] These are rounded numbers, obtained from the Dark Web Forum Portal at the University of Arizona (http://ai.arizona.edu/research/terror/, accessed 15 February 2012).

[9] It is entirely plausible that some impostors have occasionally ended up unwittingly studying each other, much like the characters in G. K. Chesterton's novel *The Man Who Was Thursday*. I am not aware of any specific such episodes, but victims are presumably unlikely to report them. It is fair to assume, though, that this has been less of a problem on the larger forums, where the total number of users is so large that the proportion of impostors must have been relatively small.

who make a point of being visible. Each of these actor categories has a different motivation for frequenting the forums. Important for our purposes is the fact that only a small subset of visitors—the operatives and some of the consumers—have an interest in engaging in a genuine recruitment exchange.

The Trust Problem

There are good reasons to expect distrust among jihadi forum users. Online trust research showed early that computer-mediated communication (CMC) is fraught with problems (Parks and Floyd 1996) and that interpersonal trust is notoriously difficult to establish in electronic contexts (Rocco 1998). The obstacles to online trust are many (Nissenbaum 2001), the most important being the absence of non-verbal cues, which gives CMC a 'narrower bandwidth' than face-to-face (FtF) communication. Experimental studies suggest a correlation between the width of a given medium's range of cue types on the one hand and interpersonal trust on the other (Bos et al. 2002). Hancock (2007) also found that highly motivated liars get away with more in CMC than in FtF. Surveys also document a lower general trust in Internet-based acquaintances and transactions than in offline ones (Naquin and Paulson 2003). Studies of romantic online relationships show that involvement tends to be lower and misrepresentation (e.g. of age and appearance) tends to be higher in cyberspace than in offline relationships (Cornwell and Lundgren 2001).

In low-risk exchanges, such as casual chatting, low bandwidth is not necessarily a big problem. On the contrary, early users of chat services argued that the Internet was 'powerfully conducive to intimacy' and that people open up sooner to anonymous online friends than they would outside cyberspace (Van Gelder 1985). In a study of online communication, Parks and Floyd (1996, 274) noted that 'personal relationships were found far more often and at a far higher level [...] than can be accounted for by the reduced-cues perspective'.

However, when stakes of an interaction increase to involve, say, the love, money, or especially the personal safety of the truster, then it becomes a different ballgame. The dangers of deceptive mimicry have been known since the early days of the Internet (Donath 1998; Van Gelder 1985; Hancock 2007; Lea and Spears 1995; Myers 1987). There was the famous case in the early 1980s of a male New York psychiatrist who passed himself off as a disabled woman named 'Joan' on CompuServe networks in an odd search for female emotional intimacy (Van Gelder 1985). Pecuniary fraud has long marred e-commerce. Around 2000, three per cent of all American consumers had experienced credit card fraud while ten per cent had paid for items online that were never delivered (Fox 2000; Williams 2001). More recently, the 'Robin Sage Experiment' (Ryan 2010), in which a security consultant used a fake Facebook profile to gather sensitive information about members of the US intelligence community and major corporations, illustrated that digital mimicry can be a national security concern.[10]

Unfortunately we know little about the sources of trust between people online, i.e. the salient variables affecting trusting decisions. The online trust literature has paid far more attention to trust in commercial websites than to trust in individuals (Corritore, Kracher, and Wiedenbeck 2003). There is a large body of research aimed at finding ways to improve consumer trust in websites (Anderson 2006; Batya and Kahn Jr 2000; Frankel 2001; Grazioli and Jarvenpaa 2000; Koehn 2003; Wang and Emurian 2005). The sources of online interpersonal trust, on the other hand, are less well understood (Donath 1999; Feng, Lazar, and Preece 2004; Green 2007; Kling 1996). There has been some work on avatars (Bailenson and Beall 2006; Galanxhi and Nah 2007), but avatar use appears to have only minor effects on perceived trustworthiness, even in low-risk interactions.

The literature has more to say about the sources of trust in offline contexts.[11] A much-cited contribution is that of Sztompka (2000), who argues that people determine the trustworthiness of others based on three main variables: reputation (record of past deeds), performance (present conduct), and appearance. It is not immediately obvious, however, how these variables are assessed in cyberspace, where information about things like appearance and performance is both limited and unreliable. Moreover, reputation may not be relevant at all when a person's very identity is in question. In fact, very often, the main problem for high-risk activists online is not that interlocutors are less reliable than they say they are, but that they simply are not who they say they are.

Such 'deceptive mimicry' has been analysed by Gambetta (2005), who argues that trusters vet trustees by looking for 'costly signs'. He proposes an analytical framework that builds on signalling theory and the idea of differential sign costs. He adds two important components, namely, a classification of signs and a taxonomy of mimicry systems. The former distinguishes between 'cues' (congenital features), 'marks' (lifestyle by-products), and

---

[10] I thank Mark Stout for drawing the Robin Sage story to my attention.
[11] See Borum (2010) for a review.

'symbolic signs' (conventional gestures, dress, or statements), with cues and marks being costlier to mimic than symbolic signs. The notion of mimicry system refers to the constellation between dupe, mimic, and model in a given mimicry episode. As the other chapters in this book demonstrate, Gambetta's framework has proved analytically useful in studies of many types of real-life mimicry in the institutional setting of civil war or armed conflict. Transposed to an online setting, however, Gambetta's classification of signs is not directly applicable, because the variety of available signs is narrower. The principle of cost discrimination probably still applies in online trust games, but the sources of the costs that discriminate are more limited. Gambetta (2005, 225) himself expects mimicry to be easier in low-bandwidth media ('in the dark or over the telephone'), but does not predict the precise effect of the low bandwidth on what he calls the 'semiotic structure of signs'.

In this chapter, I apply analytical concepts from the trust literature with minor modifications. I distinguish trust (the willingness to accept vulnerability) from trustworthiness (the individual property that inspires trust). I also reserve the term 'trust' for human interaction; things or technologies are objects of 'confidence'. In this context, confidence in communications technology is synonymous with low fear of surveillance. However, a website or a document can represent people and thus be an object of trust (i.e. I may trust a forum because I consider its administrators to be trustworthy). Although the main focus is on interpersonal trust, the chapter also considers confidence in technology, since both affect users' willingness to share sensitive information.

I focus on deceptive identity mimicry, but recognize that the threat of it is but one of several sources of interpersonal distrust. One need not be considered a potential spy to be untrustworthy; lack of commitment or sloppiness can also instil distrust. Jihadi forums contain several mimicry systems, of which I consider mainly one, namely, what Gambetta (2005) would call the 'mimic versus dupe model'. Here, spies (mimics) pass themselves off as genuine activists (models) in order to obtain sensitive information from, or confuse, the genuine activists (dupes).[12] In this system model and dupe are the same category of people.

---

[12] Another system not examined here is 'mimic versus dupe via the model' which comes in two varieties: a) senior activists from group X produce propaganda in the name of a fake group Y in order to mislead spies, and b) unaffiliated junior activists try to pass themselves off as members of a known group to get attention from other activists or to mislead spies. One might expect to see a third variant of this system in the form of activists posing as informants in order to get sensitive information from spies, but I have not observed this, at least not on the Internet. However, it happens in real life, as illustrated by the case of the Jordanian double agent who killed six CIA operatives in Afghanistan in late 2009 after posing as an informant (Warrick 2011). One might also add the phenomenon of negative mimicry or camouflage, in which activists (mimics) mislead spies (dupes) by trying to appear less radical than they really are.

Put differently, I am mainly concerned with trust in the sense that administrators and other genuine jihadi forum users have a problem deciding whether other people who join the forums are to be trusted or not as honest partners in communication. However, I shall also briefly discuss trust in the opposite direction, namely, potential users wondering whether to join a given forum given that the forum itself may be a mimic (or honeypot) designed to lure them. So everyone in this larger game can be at once a truster and a trustee.

Sources and Methods

I rely on messages collected openly from mostly Arabic-language jihadi forums, well aware that these data pose several challenges.[13] The first is an epistemological one, namely, that trust is difficult to observe in forum posts, because users rarely make the outcome of their trusting decisions explicit. Positive trust may be expressed in many different ways depending on the purpose of the conversation, and distrust may produce either suspicious remarks or nothing at all. In fact, when someone decides not to trust, nothing special happens, and we will not even know that a trust exchange occurred in the first place. The second problem regards validity: it is difficult to extract a representative sample of messages because the universe of jihadi forums is diverse and shifting.[14] Different forums may have slightly different user populations, constraints, and cultures, which presumably makes for variation in trust dynamics. Third, the reliability of the data is vulnerable to the same external manipulation that undermines online trust among jihadis. Some expressions of distrust, for example, may be forgeries produced precisely to undermine trust.

I pursued a four-pronged data collection strategy. First, I gathered a body of secondary literature on online jihadism by searching academic databases (EBSCO, WorldCat, Google Scholar) and media databases (Lexis-Nexus, Google News) for relevant terms such as: Internet/online/

---

[13] There are jihadi forums in several different languages (including English, French, Swahili, etc.), but I examined mostly Arabic-language forums because the 'market leaders' that is, the forums with the most traffic, have tended to be in Arabic. I do not know whether language matters for the level of repression or trust on the forums. It is presumably somewhat cheaper for Western security agencies to monitor and infiltrate English-language forums, because Arabic-speaking personnel are a scarce resource, but the opposite is true for security agencies of Arab countries. We should assume that most infiltration operations are carried out by native speakers, lest trustees spot mimics through language gaffes or inconsistencies.

[14] As noted earlier, I estimate that there were between five and fifteen significant jihadi forums in operation at any one time from the early 2000s to the early 2010s. A total of fifty to a hundred different jihadi forums may have seen the light of day in this period. Most had relatively little traffic and/or were short-lived.

web/cyberspace/forums AND terrorism/terrorist/jihad/jihadism/Qaida/ Qaeda. I then reviewed this literature, doing my best to identify references to: 1) evidence of surveillance and infiltration on forums; 2) cases of successful direct recruitment or fundraising through forums; and 3) forum posts expressing distrust, such as spy accusations levelled at others, general warnings against surveillance, advice on how to spot an impostor, and the like. I then dug deeper into each case by conducting additional searches and examining the available primary sources.

Second, I conducted a systematic search for select spying-related terms in the archive of the Arabic-language forum Falluja, available in a database called the Dark Web Forum Portal (DWFP) established by the University of Arizona (Chen et al. 2011). The archive covers the years 2006 through 2009 and includes content from all forum subsections, 96,000 threads in total. From these I collected all threads containing the Arabic words *tajassus* (spying), *jawasis* (spies), and *ikhtiraq* (infiltration). This yielded a sample of 314 threads. I manually coded posts by topic to generate some descriptive statistics, and I used content analysis to identify the posts with the most detailed description of sign assessment.

Third, I browsed a small sample of forum threads from the 'Umma affairs' section of the Arabic-language Ansar al-Mujahidin forum from the second half of 2011.[15] I examined the posts, looking for: 1) examples of users divulging autobiographical information; and 2) expressions of distrust of the same type as described above. My sample consisted of all threads that were last dated on the ninth and twenty-third of each month from July to November 2011, 232 in all.[16] These threads contain between one and 202 individual posts, although most contain less than ten posts. In this period, Ansar al-Mujahidin was the second most frequented Arabic-language jihadi forum.

Fourth, I examined the architecture and formal features of five forums (Falluja, Shumukh, Ansar al-Mujahidin, Fida, Jahad)—all selected for their relatively high popularity and stability—at different points in time between

---

[15] http://as-ansar.com/vb/forumdisplay.php? f=3 (accessed 16 February 2013). The 'Umma Affairs' section is a regular feature on jihadi forums and constitutes its centre of gravity; this is the most active (measured in threads per day) of all the forum sections, and it is where the latest news or the hottest topics are discussed. It arguably reflects the general atmosphere on the forum better than any of the other sections, which are devoted to more specialized topics such as legal matters, literature, statements, etc. I deliberately did not sample the section on 'technical security' because it is devoted to countersurveillance issues and would lead to overreporting of infiltration concerns.

[16] The 'last dated' criterion was primarily chosen for convenience since threads are listed that way on the forums, but it also has the advantage of adding more randomness to the sample, since there is presumably no systematic relationship between a thread's initiation date and its last modification date. A thread last dated on, say, 23 July, could conceivably have been started on any day earlier that month, or even earlier.

2009 and 2013. I looked especially for design features (such as reputation systems) that appeared to reflect trust concerns or have a trust-building purpose.

I then subjected these data to a three-step qualitative analysis. I started by reviewing open-source evidence of infiltration of jihadi forums to establish that the threat of deception is real and consequential. I then assessed the level of general trust on the forums through two indicators: 1) evidence that jihadis engage in sensitive transactions (such as recruitment) with people they first encounter on the forums; and 2) the number and content of messages expressing distrust. In the third step I sought to understand how users assessed signs in interlocutors and which strategies they use to avoid deception.

To access the forums, I myself did, to some extent, pose as a jihadist. I created user accounts under a 'jihadi-sounding' nickname to avoid eviction by forum administrators. I did not, however, use the account for anything other than observation. I never interacted with other users or posted any messages.

Assessing the Repression Level

Radical Islamists have good reason to be cautious online. Governments, hacker groups, and other activists take a range of hostile measures against jihadi websites (Shahar 2007; Weimann 2006). Much of this activity is shrouded in secrecy, but we can get a whiff of what is going on from secondary sources.

Although some governments appear to have the technical capability to at least temporarily close down jihadi websites and occasionally do so (ICT/JWMG Team 2012, 35–39; Mantel 2009, 132–135), most governments pursue a strategy of surveillance and subversion rather than direct attack. Closing websites permanently is legally complicated, users can always migrate to new websites, and there are intelligence benefits to leaving forums intact.

Jihadis thus face two main types of dangers online: surveillance and deceptive mimicry. Each involves a different set of dilemmas for its targets. Surveillance undermines confidence in the communication technology itself, while mimicry affects trust in one's interlocutors. Surveillance can be met with technological solutions (such as encryption), while deception requires vetting procedures.

Surveillance

Jihadi forums appear to be under close surveillance. Since at least the early 2000s, agencies such as the US National Security Agency (NSA) and the UK Government Communication Headquarters have had automated systems that capture, sift, and store both content and user data on a range of digital communication channels, including forums (Talbot 2005). Documents leaked in 2013 by former NSA contractor Edward Snowden suggest that the surveillance and tracking capabilities of the NSA and GCHQ had reached a formidable level by the early 2010s. In addition, both intelligence services and private monitoring companies such as SITE and MEMRI manually track messaging content on jihadi forums.[17] While formidable, the surveillance is clearly not comprehensive, otherwise all convictable online militants would presumably be captured. One challenge for authorities is stealth technology, such as web proxies, encryption software, and the like. Another is the sheer volume of traffic, which means that a particular message may be overlooked by analysts even if it is intercepted by their computers. Still, the tracking capability of governments is such that most high-profile online activists appear to get caught eventually, even if they are very computer-savvy. For example, when the prominent forum figure 'Irhabi 007' (aka Younis Tsouli) was arrested in London in October 2005, it was considered remarkable that he had been able to taunt his online pursuers for two whole years before being caught.

Deceptive Mimicry

Jihadis face three main types of deceptive mimicry on the forums: fake websites, false statements, and impersonations. The most famous case of website mimicry is that of the al-Hisba forum, which reportedly was operated by the CIA and Saudi intelligence as a 'honeypot' from around 2006 to 2008 (Nakashima 2010).[18] Officials said the operation had been 'a boon to Saudi intelligence operatives, who were able to round up some extremists before

---

[17] The Search for International Terrorist Entitities (SITE) Institute is a for-profit company specializing in the monitoring of online terrorist propaganda. The Middle East Media Research Institute (MEMRI) is a not-for-profit media monitoring organization. See https://news.siteintelgroup.com/ and www.memri. org for details.

[18] The forum was eventually dismantled in 2008 by the US military, reportedly against the wishes of the CIA. The former saw the forum as helping insurgents in Iraq, while the latter saw it as a useful intelligence source (Nakashima 2010).

they could strike' (Nakashima 2010). A similar operation may have been attempted in 2009, when the al-Ekhlaas forum mysteriously resurfaced after a year's absence, at the same time as the al-Fajr media distribution network was reportedly hacked and used to encourage forum members to sign up for al-Ekhlaas (Rawnsley 2011). Dutch intelligence (AIVD) is also reported to have carried out a honeypot operation, the details of which are not publicly known (Torres Soriano 2012).

There is also some evidence that intelligence agencies manipulate statements in the name of jihadi organizations, though this is less well documented. Schmitt and Shanker (2011) quote a US intelligence official as saying that the ability of the intelligence community to persuasively imitate Al Qaeda web postings 'does give an opportunity for confusion [...] if we can post an almost-authentic message. We have learned to mimic their "watermarks"'. The official did not indicate which particular type of statements they had imitated. Another type of document manipulation occurs when authorities capture unpublished propaganda during police raids and then post the material strategically on jihadi forums to see who accesses it. Saudi intelligence appears to have conducted such a 'mini-honeypot' operation against 'al-Qaida on the Arabian Peninsula' in early 2006 (Ulph 2006a).

Yet another variant of forgery is when governments distribute obviously false documents to send a blunt message that they are watching. In 2004, for example, online jihadists were confused by the appearance of two different versions of issue 14 of *Sawt al-Jihad* magazine, one of which appears to have been a government forgery (Torres Soriano 2012). More entertaining was the so-called 'Operation Cupcake' in 2010, in which British intelligence substituted bomb-making instructions in the magazine *Inspire* with a cupcake recipe (Gardham 2011). The purpose was primarily to make the bomb recipe unavailable, but perhaps also to poke fun at the real authors. Having said all this, it is implausible that governments manipulate more than a small proportion of all statements on the forums, because the overall quantity of propaganda is extremely large, and credible forgeries require highly skilled labour.

The third and for our purposes most important type of mimicry involves agents posing as real jihadis. According to Schmitt and Shanker (2011), the US military has 'Digital Engagement Teams' consisting of fluent speakers of Arabic and other languages who infiltrate the forums under fake identities and 'over time build up credibility and sow confusion'. There are also unconfirmed but highly plausible claims that intelligence services from Arab countries, notably Jordan and Saudi Arabia, do the same (Ulph 2006b).

It is worth pointing out that, for government agencies fighting jihadism, forum infiltration involves two partially contradictory objectives. One is to discover information, and for this it is helpful to have mimics succeed. The other is to spread distrust among jihadists, in which case it may be advantageous to have at least some of the mimics fail and be discovered in order to sow suspicion and encourage self-censorship. The contradiction goes deeper: if the latter strategy succeeds, then less valuable information will be forthcoming as people will be more prudent or stay out altogether. Thus, if agencies think that valuable information can be acquired via forums then the latter strategy is counterproductive. If, by contrast, they think that little of value can come from the forums, then why bother with either strategy? This suggests that agencies may have an interest in not letting trust break down completely on the forums.

Detailed accounts of deceptive mimicry episodes are rare, presumably because victims are imprisoned and governments are protective of their methods. It could also be that successful cases of deception are relatively infrequent due to the precautions taken by activists; we simply do not know. However, they do happen, and a good account is found in the testimony of Shannen Rossmiller, a municipal court judge in Montana who infiltrated forums as a self-appointed cyber-vigilante in the early 2000s (Rossmiller 2007). This is her account of the interaction with Ryan Anderson, a US soldier and convert to Islam:

In October [2003], while monitoring Arabic Islamist websites [...], I saw a message posted in English by a man calling himself Amir Abdul Rashid. He said he was a Muslim convert who was in a position to take things to the next level in the fight against our enemy (the US government). He further requested that someone from the mujahideen contact him for details. I was suspicious because Rashid posted his message in English on an Arabic website and was openly seeking contact from the mujahideen. I traced his IP address back to an area outside of Seattle, Washington. Over time, it also became apparent to me that he was a member of the US military. I posed as an Algerian with ties to that country's Armed Islamic Group [...], and sent Rashid an email in English with the subject line 'A Call to Jihad'. Rashid responded by asking if it was possible that a brother fighting on the wrong side could sign up or defect, so to speak. Over a period of four months, Rashid and I exchanged a series of thirty emails in English. I learned he was a member of the Army National Guard from Washington State, whose tank battalion unit was scheduled to be deployed to Iraq in February 2004. Through the course of our email exchanges, Rashid provided me with information and materials on the weaknesses and vulnerabilities of the M1-Al and M1-A2 Abrams tanks as well as US troop locations in Iraq. At all times during our communications, Rashid perceived me to be a mid-level Al-Qaeda operative. After our fifth email exchange, I contacted the Department of Homeland Security, which put me in contact with my local FBI office.

Anderson's email exchange with Rossmiller earned him life in prison, which shows that bad trusting decisions on jihadi forums can carry very high costs. There are numerous other examples of people having been imprisoned as a result of their online activities (Mantel 2009). In addition to hurting the individual, deception can damage networks and organizations. Schmitt and Shanker (2011) report one confirmed case in which 'a jihadist website was hacked by American cyberwarriors to lure a high-value Al Qaeda leader to a surreptitious meeting with extremist counterparts only to find a US military team in waiting'.

The policing of jihadi forums appears to have become more systematic and aggressive over time. In 2001, the US Patriot Act made it illegal to advise or assist terrorists, including via the Internet, while the UK Terrorism Act of 2006 criminalized the encouragement or glorification of terrorism, including through Internet messaging. The mid-2000s saw a marked growth in forum monitoring and analysis by private companies, academics, and other observers.[19] It is reasonable to assume that a similar, if not larger, monitoring effort was exerted in the intelligence community.

Gauging Trust Levels

Given what we know about the fragility of online trust and the dangers lurking on jihadi forums, we can make two testable predictions. First, we should expect users to be reluctant to engage in sensitive transactions that require divulging real-life contact details, because this makes them vulnerable to apprehension. Second, we should expect users to verbally express concerns about surveillance and deceptive mimicry.

Sensitive Transactions

The available data suggest that sensitive transactions, which I assume to be indicators of positive trust, were very rare on the forums in the time period studied. Such transactions probably did happen occasionally, but if they had

---

[19] Witness the increase in the output of monitoring services (e.g. SITE, MEMRI, BBC Monitoring, and World News Connection), journals (e.g. *Jamestown Terrorism Focus*), and blogs (e.g. Internet Haganah) devoted to online jihadism during the late 2000s.

been very widespread we would arguably have seen more traces of them in the sources.[20]

In fact, in the messaging data collected for this chapter, I observed no cases of direct recruitment, fundraising, or operational coordination mediated through the forums. There were no examples of two users exchanging contact details that would have allowed them to meet in real life. Nor were there cases of people exchanging account details that would have allowed for money transfers. Of course, I only reviewed a few thousand messages, so this is not sufficient evidence to conclude. Moreover, as we know that some forums had restricted sections (access to which was limited to certain users), it is possible that sensitive transactions took place in spaces I could not observe. However, I did not observe any cases of people meeting in an open section and agreeing to meet virtually in a restricted section.

This negative finding would be less credible were it not for the fact that other studies also report low levels of direct recruitment (Rogan 2006b; Sageman 2004; Weimann 2007), operational coordination (Bergin et al. 2009), and financial transactions (Policy Planners' Network 2011) mediated directly through the forums. Moreover, one of the few available in-depth case studies of the recruitment tactics of a jihadi organization found that al-Qaida in Saudi Arabia did not recruit directly online (Hegghammer 2013a).

To be sure, some studies do argue that jihadis recruit, plot, and fundraise online (Denning 2010; Theohary and Rollins 2011). They point to the fact that forums contain recruitment and fundraising calls, as well as travel advice for jihad fronts such as Iraq and Afghanistan (Kohlmann 2008). They also point to the several cases of people who say they joined a jihadi group after reading propaganda on jihadi forums.

However, these studies tend to speak of recruitment or fundraising in a broad sense, conflating radicalization processes inspired by online propaganda and recruitment transactions executed in cyberspace. On closer inspection, most of the available evidence of online recruitment are cases of forum-*inspired*, not forum-*mediated* recruitment. In most examples of online recruitment—see ICT/JWMG (2012) for a good overview—we only hear about the recruiting organization reaching out (the demand side) or about the recruit responding to a general call (the supply side). None of the examples are cases of recruiters and recruits meeting online and exchanging

[20] The most prominent possible case of operational forum use was reported in August 2014, when the *Daily Beast* quoted intelligence sources saying that the top al-Qaida leadership had convened some kind of online meeting with representatives of several al-Qaida affiliates (Cruickshank and Lister 2013). An anonymous intelligence official told *Associated Press* 'the threat began with a message from al-Wahishi ... to al-Zawahri ... The message essentially sought out al-Zawahri's blessing to launch attacks. Al-Zawahri, in turn, sent out a response that was shared on the secretive online jihadi forum' (Jakes and Goldman 2013). However, other anonymous officials described the location of the meeting simply as 'a virtual meeting space'. To this day it is still not clear whether the communication took place on a jihadi forum.

contact details that enabled them to meet in real life. Forums probably do affect recruits' wish to be recruited or funders' wish to donate money, but they appear not to mediate sensitive transactions between people who only know each other online, at least not on a large scale.

A sceptic might ask why we should expect jihadists to exchange contact details on a forum in the first place, and whether the absence of such transactions tells us anything about trust. To this I would respond that al-Qaida and its affiliates are under-resourced and that forums have the potential to provide access to a much larger pool of recruits and funders than they would otherwise reach. For a labour market as small as this, where willing recruits are thinly spread out across the globe, the Internet could prove a highly effective mechanism connecting supply and demand. In an imaginary world where jihadis could operate freely, forums could serve as the LinkedIn or Craigslist for terrorist labour. For a more pertinent baseline comparison we may look to less contentious transnational activists such as WTO and G8 summit protesters, who *have* been able to use social media for recruitment and operational coordination (Van Lear and Van Aelst 2009, 236ff; Tarrow 2005).

Expressions of Distrust

In the absence of other indicators of positive trust, we may look to expressions of distrust as a negative indicator. Presumably, the higher the frequency and intensity of the expressions of distrust, the lower the level of general trust in a community.

As one might expect, jihadi forums contain many warnings against spies, rumours of infiltration, advice on stealth tactics, and the like. To be sure, expressions of distrust represent only a small proportion of all forum posts. According to my Falluja message count, only some 0.3 per cent of all threads (314 of 96,000) contained the words 'spying', 'spies', or 'infiltration'. Considering that this is only a subset of all possible words of distrust, my figure is reasonably consistent with the finding of another study by Aaron Weisburd, which found some 1.5 per cent of threads to be devoted to countersurveillance.[21] This may not seem like much, but it is sufficient for the issue to come to readers' attention on a regular basis. Moreover, such messages may attract disproportionate attention; for example, on the Ummah News section on Ansar al-Mujahideen English forum in November 2011, three out of

---

[21] Author's email correspondence with Aaron Weisburd, 23 February 2011. The study has not been published.

nine sticky posts were about how to avoid surveillance and spies on the forums.[22]

Another metric worth mentioning is the extent to which activists use web proxies, which hide IP addresses, or https URLs, which protect against data interception. According to Aaron Weisburd of Internet-Haganah.com, around ten to twelve per cent of jihadi forum users in 2010 employed proxies while even fewer used https URLs.[23] These relatively low rates may be partly explained by the fact that most visitors only access forums to consume propaganda, a relatively risk-free activity.

The concerns expressed in forum messages address a range of issues.[24] Broadly speaking, they seem to cluster around the four dangers mentioned above, namely: 1) surveillance; 2) website hijacking; 3) propaganda forgery; and 4) fake identities. In the following I present examples of such expressions.

First are surveillance fears, which make up the bulk of distrust messaging on the forums. Recurrent topics include:

- general warnings about surveillance[25]
- warnings about surveillance on other forums and communication channels (Biyokulule 2009)
- warnings against the use of internet cafes (Ulph 2006c)
- warnings against viruses and spyware (Rawnsley 2011)
- advice on using TOR and other advanced proxies[26]
- advice on encryption software (Vijayan 2008)
- warnings about fake encryption software (Rawnsley 2011)

While most concerns appear in the form of separate threads or documents, they are also expressed during online conversations or in private messages. For example, in 2003 a user nicknamed 'Aqil al-Masri' actively sought travel advice for Iraq on jihadi forums, after which he was contacted by another forum user who warned him against posting so much personal information online.[27] Another academic forum observer, Torres Soriano (2012), has noted jihadi 'mistrust of anything originating in cyberspace: antivirus

---

[22] See http://www.ansar1.info/showthread.php?t=31326; http://www.ansar1.info/showthread.php?t=33,459; and http://www.ansar1.info/showthread.php?t=29715 (accessed 21 November 2011).
[23] Author's email correspondence with Aaron Weisburd, 23 February 2011.
[24] For an extensive list of examples, see ICT/JWMG (2012).
[25] See, e.g. 'Essential advice and guidance to users of jihadi forums', http://www.as-ansar.com/vb/showthread.php?t=31308, dated 1 December 2010 (accessed 21 November 2011).
[26] 'Explanation of how to use TOR, and how security and intelligence agencies can catch you like a fish', http://202.71.102.68/alfaloj/vb/showthread.php?t=46,046 (accessed 9 August 2009).
[27] http://www.al-ekhlaas.net/forum/showthread.php?t=50742 (accessed 12 November 2007). I thank Truls Tønnessen for sharing this document.

software, navigation tools, commercial email accounts, and so on. Absolutely everything is suspected of concealing a trap. For example, a Jihadist site warned its followers "to be careful with Google".

Second, forum users periodically flag their suspicion that specific forums and websites are operated or infiltrated by intelligence agents. Some examples:

- In 2004, the jihadist forum Al-Ma'sada warned about a possible CIA-sponsored forum called Batal (Ulph 2004).
- In early 2006, the Hisba forum was widely accused of being a 'den of spies' by users on other forums, notably Tajdeed (Ulph 2006c; Weisburd 2006a).
- In November 2009, a Shumukh forum user issued a 'Warning to followers and pioneers of jihadi forums about the Electronic Mujahidin Network'.[28]
- In 2010, an alleged Taliban spokesman warned on Falluja that 'the [Taliban] main site and the site of its online journal al-Sumud, have been the subject of an "infiltration operation"'. It warned others 'to not enter any of the links that concern these websites, and not even to surf [the content] until you receive the confirmed news by your brothers' (Rawnsley 2010).
- In mid-2010, following a temporary, near-simultaneous shutdown of several jihadi forums, 'sites began blaming one another for the forums' collapse, with accusations about collaborating with the enemy against rival sites, or of having been infiltrated by enemy agents. Forum participants wondered which sites could be trusted and which were recognized by Al-Fajr, the prestigious jihadi media company organization that distributes videos and communiqués from Al-Qaeda and its franchises' (Hazan 2010).

Torres Soriano (2012) argues that the forum users' fear of infiltration has reached the stage of paranoia and notes that 'due to the fear of alleged infiltration, apparently innocuous events take on conspiratorial tones'.

A third issue of concern is the authenticity of specific statements and videos. However, this suspicion is expressed much less frequently than surveillance or infiltration fears. One such message was entitled 'A warning for those who post unconfirmed and false news about Jihad and Mujahideen'.[29]

---

[28] http://www.shamikh1.info/vb/showthread.php?t=52283 (accessed 21 November 2011).
[29] http://www.ansar1.info/showthread.php?t=33459 (21 accessed November 2011).

Fourth and finally, numerous forum activists worry about agent provocateurs. Some posts advise general precaution, such as: 'do not use the Internet for recruitment of new members under any circumstances'.[30] Others seek to intimidate potential spies, as did a 'Letter to the Treacherous Forum Spies' in 2011.[31] At other times, accusations target specific users. For example, in January 2004, the famous activist named 'Irhabi007' wrote: 'My brothers, I am convinced that there is a person spying on the forum, but fortunately I have figured out his IP, which is 62.233.243.116. The IP is registered in Poland' (Weisburd 2006b; 2006a). In some rare instances, victims of hacking warned about fake messages being sent from their own account; for example, in August 2011 a Shumukh user wrote: 'it seems that someone is using my account and is somehow sending messages with my name to the members' (Rawnsley 2011).

Forums also carry accusations against offline spies. In 2006, for example, there was much stir on the forums about a certain Abu al-Qaʿqaʿ, an alleged Syrian agent supposedly involved in the capture of several volunteer fighters on the way to Iraq (Moubayed 2006). In August 2009 the name, picture, phone number, and other details of an alleged Jordanian agent were posted on the Shumukh forum.[32] Forums have also been used to spread ideological treatises on spying in general and how to punish it. At least three such works have been widely publicized on the forums in recent years, namely: Abu Yahya al-Libi's (2009) 'Guidance on the Ruling on the Muslim Spy'; Abu al-Nur al-Maqdisi's (2009) 'Ruling on the Spy in Islam'; and a more practical text entitled 'Dealing with Spies' by a certain Sayf al-Jihad (2010). Although these warnings and treatises are not expressions of *online* distrust, they are suggestive of a general fear of infiltration.

It is not surprising, then, that a substantial number of posts conclude that forums should not, under any circumstances, be used to exchange sensitive information. For example, one message on the Falluja forum in 2008 cautioned (ICT/JWMG Team 2012):

> Jihadist Web forums are not the appropriate place to plan or coordinate action in the field—that is, to plan an attack, to recruit participants in an attack, or to discuss implementing an attack, whether in a surfer's home country or abroad. The forums are not sufficiently secure for this purpose: as noted, intelligence services

---

[30] 911 Ghayr Muttasal, '*al-ʿamal al-jihadi al-sirri: istratijiyat* (Strategies of secret jihadi operations)' *Muntadayat al-Maʿsada al-Jihadiya*, 18 March 2005 (accessed 13 May2005).

[31] http://sfir-arabicsource.blogspot.com/2008/06/rant-against-spies-in-forums.html (accessed 21 November 2011).

[32] http://www.shamikh1.info/vb/showthread.php?p=252283 (accessed 21 November 2011).

can hack into them and their databases. Jihadist Web forums are not the place to make contact with other surfers or to divulge personal details—even if the other surfer asking for them has previously instructed you in the ways of jihad.

It should be noted that some qualified observers have assessed forum trust levels differently. A study by Dutch intelligence (AIVD 2012, 13) argued that: 'there is great mutual trust on core forums [...] The virtual trust among members of such networks can be so unconditional that they may decide to meet offline and discover each other's true identity'. The study offered no examples of such meetings or estimates of their frequency. It is difficult to reconcile their assessment with mine. Perhaps classified sources show a different picture than open ones; perhaps AIVD sampled a forum where trust levels happen to be higher, or perhaps their statement is a generalization based on a small number of cases.

Interestingly, my data indicate that the level of distrust on the forums appears to have increased over time, especially in the second half of the 2000s. This author recalls seeing many cases in the early 2000s of forum users volunteering sensitive information such as their geographical location or even family connections.[33] For example, Saudi forum members would often post the telephone numbers to the homes of fellow Saudis killed in Iraq, indicating that they knew these families personally (Kohlmann 2008). Several long-time academic forum watchers have reported the same impression. Labi (2006) noted:

In the early days, before the Iraq War, the online global jihad amounted to a collection of chat rooms where angry members could let off steam and experiment with threatening graphics. The sites welcomed visitors, offering a painless process of registration; today they present tougher barriers to entry and place a greater emphasis on remaining anonymous and secure.

Similarly, Ulph (2006c) wrote that:

the number of postings on the Internet on the subject of infiltration has lately increased. Each jihadi forum site already includes a section on the use of proxies, giving detailed instructions on how to disguise the identity of reader and contributor and practical security precautions to take to ensure identity security.

Likewise, Shahar (2007) observed:

Jihadists' confidence in their ability to dodge state control via the use of Internet forums has dropped significantly in the past couple of years. Despite the jihadist

---

[33] Unfortunately I did not think of storing the data at the time, so I cannot document these cases.

forum administrators' best efforts to use proxies and to conceal participants' identities, this kind of confidence may not be all that easily recovered.

By 2011, the forum users had become so jittery, according to one observer, that they 'panic on hair trigger alert, at times blowing cyber incidents out of proportion' (Rawnsley 2011).

My count of spy-related threads on Falluja between 2006 and 2010 partly corroborates this anecdotal evidence. The data suggest that interest in spying as a general topic (i.e. offline as well as online) was stable in the late 2000s. However, the proportion of spy threads pertaining to online matters increased steadily, from eleven per cent in 2006 to twenty-two per cent in 2009. The volume of messaging increased significantly in the same period.

It is reasonable to assume that the increased interest in electronic spying is at least partly a response to the increase in government surveillance and subversion of jihadi forums from the mid-2000s onward. Increased repression, in other words, seems to have exacerbated the trust problem.

Sign Knowledge on Jihadi Forums

Exactly which signs make or break trust? Knowing how jihadi forum users assess the trustworthiness of individual interlocutors is very difficult, since we have no detailed first-person accounts of trust exchanges. What we do have, however, is a number of texts offering general advice on how to spot a spy, a genre that developed, perhaps not coincidentally, in the late 2000s. In principle, if we know what users consider red flags, we might be able to infer what they consider signs of trustworthiness. Let us take a look at two such texts.

The first is a short forum post from 2009 entitled 'Fake Profiles on Jihadi Networks and Forums' (Abu Jarir 2009) which lists six suspicious behaviours:

- 'Working to sow discord among the mujahidin and their supporters'
- 'Casting doubt on the overall method of the mujahidin, their leaders and clerics by highlighting their errors and disputes'
- 'Working to hold Muslims back from jihad and supporting the mujahidin'
- 'Asking about things regarding the mujahidin on the battlefronts, to obtain information enabling strikes at them' [...] 'for example, one of them wrote asking about the number of groups fighting in Iraq, and are they mostly Sunni or Shiite? In which areas are they based? What are their names?'

- 'Indirectly tarnishing the image of the mujahidin by mentioning some of their criminal acts such as killing ordinary Muslims and blowing up cars in markets, etc.'
- 'Working to spread defeatism and despair among members and visitors by exaggerating the capabilities of the enemy.'

Notice that these guidelines focus exclusively on verbal statements; more specifically, on the political content of these statements. By the author's logic, anyone who asks detailed or critical questions is a potential spy. As a tool for identifying fake profiles it is wholly inadequate, because it would not spot infiltrators who do not rock the boat. The only thing it can achieve is the screening out of the ideologically uncommitted and verbally undisciplined. One might speculate that the purpose of the text is in fact not to help detect infiltrators, but to exploit the threat of infiltration to weed out doubters and critical thinkers. It could be that a world full of mimics gives new opportunities to models, in this case to purge their ranks of the not so faithful.

Another, more interesting document surfaced in late 2010 under the title '10 Methods to Detect and Foil the Plots of Spies' (Anonymous 2010). In the introduction, the author makes explicit reference to a case in the US state of Oregon a few months earlier, in which a young Somali-American was arrested after being drawn into a fictitious terrorist plot staged by a FBI agent provocateur on a jihadi forum. The text was widely circulated and remained a sticky post on Ansar al-Mujahideen English forum over a year after its publication, which is much longer than usual for sticky posts. The text mentions the following red flags:

- Small talk ('Once they make contact, the dialogue is typically small. Such topics as translations, finding nasheeds, looking for a husband/wife, or best places for halal food are discussed, simply because they are low key.')
- Bluntness ('talking about Jihad from day one raises red flags')
- Inconsistencies and changes of stories ('Watch for small things which may seem insignificant, such as the mentions of family members, a job, or knowledge of a particular topic')
- Status claims ('These typically include claims to be a member of a [jihadi] organization, or to be in contact with Mujahidin [...] Anyone who makes these claims is either a liar, or [...] extremely [ignorant] when it comes to security').

- Absences ('A spy will typically claim to be busy, most likely with school [...] However, if one really seeks Jihad, and in particular [martyrdom], it is unlikely their primary focus is schooling.')
- Requests for clarification ('For example, if you say you are interested in Jihad, they will ask if you mean physical Jihad [...] because if and when you are arrested, they want to make sure a jury will understand what you meant and thus convict you.')
- Tolerance ('Typical behaviors such as smoking, listening to music, or hanging photos are never condemned [...] Spies never appear to get mad at or disagree with their targets')
- Excessive talk and training ('Most operations [...] does not need months upon months of training, but spies will make it appear as though it does [...] there will be numerous dry runs or explosive demonstration.')
- Requests for incriminating favors ('Ask yourself why do they want you to plant the bomb? [...] spies will more than likely ask to train on your property and to use your firearms if you have any.')

We can make two interesting observations here. One is that several pairs of indicators are mutually contradictory: small talk versus bluntness, requests for clarifications versus tolerance, bluntness versus excessive talk and training. On the whole, the list is replete with Catch 22s and therefore likely to instil suspicion of practically any interlocutor. The other is that all except one of these indicators are verbal cues. The exception is 'absences', which suggests that extensive online presence (i.e. high time expenditure) is one of the few non-verbal signs available to online trusters. As we shall see, time expenditure appears to be an important discriminating sign for radicals online. In addition to being a sign of willingness to incur cost for the cause, it is a sign of tenacity, consistency, and perhaps even of an obsessive mindset, qualities presumably appreciated in jihadi circles.

Strategies for Building Trust

Given that jihadi forums have existed for a while, it is fair to assume that activists have sought to mitigate the trust problem in some way. Since we have no primary sources detailing conscious efforts of this kind, I return now briefly to the online trust literature for ideas on what we might expect activists to do, before comparing these expectations with what we observe on the forums. This deductive approach will not uncover the full range of strategies used, but it will allow us to confirm whether or not a limited number of expected behaviours occur. Note also that in this section, the distinction between trust in people and confidence in things becomes somewhat blurred, because strategies used to assess the true identity of people overlap with strategies to assess the authenticity of documents.

As noted earlier, Sztompka (2000) suggests that reputation, performance, and appearance are key variables affecting trusting decisions. We should thus expect activists to do things that convey more and better information about each of these three variables in a given user or group.

At the macro level, we might expect collective measures in the form of technical or organizational systems that facilitate particular information displays. From the literature on trust in commercial exchanges we know of two systemic measures that have proved particularly effective in reducing trust problems, namely, reputation systems and vetting agencies. A reputation system preserves and conveys information about the past online behaviour of a given user account (Corritore, Kracher, and Wiedenbeck 2003). Online second-hand trading sites such as Ebay and Amazon use reputation systems with success. A vetting agency is an independent institution that approves products or services on a range of different websites. In regular online commerce, intermediaries such as Visa or Paypal can be effective mechanisms that reduce the costs and risks of personal trusting (Frankel 2009). One could imagine similar systems in the online jihadi world.

At the micro level, we should expect individual trustees to show a preference for displaying particular types of information. From Gambetta (2005) we learned that trusters vet interlocutors by looking for costly signs. While there are fewer sign types in cyberspace that reliably convey that a cost has been incurred, there are at least three types of signs that allow for some cost discrimination. One is time expenditure: spending large amounts of time on a forum is costlier than dropping by every now and then. Another is complex format use: it requires more effort to forge or manipulate a video than a written statement. A third is complex knowledge displays: certain types of insider knowledge are hard to acquire without engaging in activity that is costly in itself.

Based on these reflections, I expect to see at least four strategies: 1) the use of reputation systems; 2) the use of vetting agencies; 3) a preference for complex media formats; and 4) displays of complex knowledge. The hypothetical emphasis on time expenditure would be confirmed by the existence of reputation systems—time expenditure being a sign in a set of signs that form reputation—and is thus not treated separately.

Reputation Systems

All the five forums whose design features I examined (Falluja, Shumukh, Ansar al-Mujahidin, Fida, Jahad) had reputation systems integrated into the website architecture. Any message on these forums would be accompanied by a small display of key information about the author's history on the forum. The display typically showed: 1) the number of messages posted by the user; 2) the date of joining; and 3) one or more status titles determined by some combination of the quantity and quality of the user's past messaging. The status title tended to be displayed just below the profile name and be printed in colour (see Figure 3.2). For example, the 2012 market leader, Shumukh, had an elaborate title system, with one set of titles reflecting the quantity of messages (much like frequent flyer statuses in the airline industry), and other sets reflecting other contributions. Thus a given user would typically carry one of the six following titles:[34]

- 'Limitless member' (*shamikh bila hudud*), reserved for users who have contributed over 6000 messages.
- 'Gold member' (*shamikh dhahabi*), reserved for users who have contributed 3000–6000 messages.

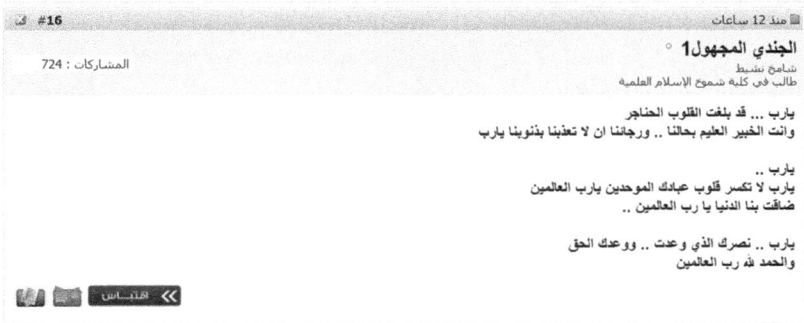

Figure 3.2 Screenshot of message from user nicknamed *al-jundi al-majhul* ('The Unknown Soldier') on the *al-Shumukh* forum, 5 February 2012. The signature is on the top right and the past message count (724) on the top left. The signature includes the titles 'Active member' and 'Student in the Shumukh Media College'

---

[34] I reconstructed the message number thresholds by compiling some 100 user profiles and examining the message numbers associated with each status level.

- 'Exclusive member' (*shamikh mumayyiz*), reserved for users who have contributed 1000–3000 messages.
- 'Active member' (*shamikh nashit*), reserved for users who have contributed 500–1000 messages.
- 'Enthusiastic member' (*shamikh muharridh*), reserved for users who have contributed 50–500 messages.
- 'New member' (*shamikh jadid*), reserved for users who have contributed less than fifty messages.

In addition, some users would also carry titles that presumably required attribution by some authority, such as:

- 'Administrator' (*idari shabakat shumukh al-islam*, in bright red), presumably reserved for a small number of users with forum administrator rights.
- 'Discussion Supervisor' (*majhud mumayyiz fi mutaba'at mawadi' al-shabaka*, in bold dark red), presumably a type of junior forum administrator.

Moreover, some had titles that may have been self-attributed, such as:

- 'Student in the Shumukh Media College' (*talib fi kulliyat shumukh al-islam li'l-i'lam*), a more frequent title than that of discussion supervisor; perhaps reserved for aspiring administrators.
- 'Your Sister in God' (*ukhtkum fi'llah*), title accompanying most users with female names.

Vetting Agencies

An interesting feature of the online jihadi propaganda system is the existence of entities referred to in the literature as 'distribution companies' (Kimmage 2008; Lia 2006). These entities are effectively vetting agencies that serve as intermediaries between groups in the field and the online forums. They receive original productions from established militant groups or ideologues, verify the documents before distributing them in the forums branded by their own logos (see Figure 3.3 and Figure 3.1, upper right hand side). Much like record companies, each distribution company appears to maintain a portfolio of groups to whose publications it has exclusive 'rights'. Distribution

Figure 3.3 Sample logos of jihadi distribution and production companies, February 2012

companies may occasionally distribute amateur productions by unknown 'artists'. In 2012 the most prominent distribution companies were al-Fajr Media and the Global Islamic Media Front, but there were several others.

Distribution companies are one of two key components of the propaganda distribution chain, the other being the production companies, which are responsible for the propaganda production of individual groups. Together these make up the category Daniel Kimmage (2008) calls 'MPDEs'—Media production and distribution entities—which, he argues, systematically brand jihadi media. A US intelligence official told Schmitt and Shanker (2011, 147) that 'all Al Qaeda products appear to go through a chain of preparation and approval. […] it's a complex system of validation. This makes messages posted on these sites official'. Kimmage also suggests that the current distribution system was inspired by a jihadi policy paper written in 2006 by an anonymous strategist who recommended branding as a solution to the problem of information saturation on the forums (Kimmage 2008). The policy paper does not explicitly mention trust concerns as a motivation, but the system likely increases confidence in those propaganda products that have been branded or vetted by the right agencies. It is not clear what effect, if any, these vetting agencies have on interpersonal trust on the forums, other than lending credibility to the individuals working directly for the agencies. It is worth noting that due diligence agencies specializing in the vetting of new collaborators, common in the business world, have yet to emerge on the forums.

Using Hard-to-fake Formats

While all digital messages can be manipulated, it is somewhat harder to forge documents in 'higher-bandwidth' formats such as audio or video than it is to forge a typed text. For example, in the early 2000s, when people wondered whether Usama Bin Ladin was alive, the appearance of a video statement featuring the al-Qaida leader was considered more reliable evidence than a written statement signed in his name.

Over the past ten years, there has been a dramatic increase in the quantity and sophistication of audiovisual propaganda circulating on the forums. Although hard to measure, the proportion of audiovisual documents relative to the overall propaganda production also seems to have increased. The growth in video production seems at least partly driven by trust concerns. To be sure, some of the increase observed in recent years may be caused by other factors, such as the expectation that visual messaging has a higher emotional impact on prospective recruits. However, large-n studies of jihadi video content indicate that most jihadi videos are not recruitment films, but short clips documenting individual attacks (Finsnes 2010). The purpose of these clips appears to be to demonstrate effectiveness and attract funding from sympathizers abroad through pre-existing financial channels. In this genre, credibility is presumably a more important attribute than emotional impact, so we cannot attribute the whole increase in video production to the propagandists' desire to impact emotions.

Anecdotal evidence also suggests that activists prefer complex formats when they make particularly risky trusting decisions or when they need to persuade highly sceptical interlocutors. For example, in the few known cases where representatives of militant groups have met 'live' (i.e. in real time) with anonymous recruits online, the meetings took place on the audio chat programme Paltalk and not on the forums (although their occurrence was often announced on the forums). Similarly, when the Yemeni–American jihadist ideologue Anwar al-Awlaki needed to convince his followers in late 2009 that he had survived a widely publicized US drone attack in Yemen, he chose to record an audio message and send it by attachment to supporters with whom he had until then communicated primarily by encrypted email (Swann 2011).

Displaying Hard-to-earn Knowledge

The forums are replete with displays of hard-to-acquire knowledge, most of which is either technical or cultural. It is not surprising, of course, that jihadi forums contain manuals on explosives production, advice on countersurveillance, and the like. Many forums have devoted special sections to these types of topics. Users who combine displays of high technical expertise with lengthy online presence appear to gain in status and perceived trustworthiness. Irhabi007, for example, became so respected as a tech whiz in the online community that real-life operatives such as Mirsad Bektasevic (head of a terror cell arrested in Bosnia in 2006) and Abu Maysara al-Iraqi (a senior media official of al-Qaida in Iraq) trusted him enough to engage in bilateral email correspondence.

More unexpected is the extent to which forum participants seem to care about poetry, religious hymns (anashid), and other elements of what we may call 'jihad culture' (Hegghammer 2013a). Several forums have special sections devoted to such products, which is surprising given their limited military utility. There may be several reasons for this phenomenon, one of which is that cultural knowledge is considered a proxy for time spent in the underground. It is worth noting that several of the most senior figures in the Shumukh reputation hierarchy, such as the forum administrator Abu Dharr al-Makki, include poetry in their 'signatures'.[35]

Displays of hard-to-earn knowledge seem to be valued more than simple declarations of extreme intent. Just as the author of '10 Methods' dismissed bluntness as a red flag, many forum users seem unimpressed by radical talk, presumably because it is cheap. Of course, what counts as sorting knowledge probably changes over time as frequent exposure turns rare knowledge into common knowledge. A written poem, for example, can easily be copied and pasted, so the recital of an old classic will likely not instil trust.

Conclusion

This chapter probed the effects of the trust problem on radical Islamist discussion forums, a supposed recruiting ground for al-Qaida and related groups. My data indicated that in the late 2000s, forums rarely mediated sensitive transactions such as direct recruitment. The social atmosphere on jihadi forums was characterized by distrust and paranoia: users withheld personal information, warned against intelligence agents, accused each other of spying, and shared advice on how to spot impostors. Moreover, interpersonal trust seems to have declined over time, from low levels in the mid-2000s to very low levels in the early 2010s, most likely as a result of increased government infiltration and surveillance.

Still, jihadi forums remained popular well into the 2010s when they were superseded by social media such as Twitter and Telegram. One reason for their survival was probably that propaganda consumption and the expression of radical views were generally not criminalized, so activists could do this

---

[35] http://www.shamikh1.info/vb/showthread.php?t=137267 (accessed 25 November 2011).

without risking arrest. Another reason may be that user confidence in the authenticity of propaganda products remained high, due in part to the emergence of jihadi vetting institutions and the growing availability of hard-to-fake video formats. A modicum of interpersonal trust also appeared to remain on the forums, thanks to reputation systems and the existence of a few relatively reliable signs such as spending much time online and displaying complex knowledge.

These findings highlight the trust problem as an important limit to the Internet's usefulness for high-risk activists. Although the inquiry was narrow in its empirical scope, it is fair to assume that the problem plagues most contentious activists online to a greater or lesser extent. This idea is not new—Diani (2000) suggested early that 'engaging in high-risk activities requires a level of trust [...] which is unlikely to develop if not supported by face-to-face interaction'—but we now have empirics to support Diani's prediction.

The chapter also offers a partial answer to the question of what constitutes a costly sign in cyberspace. *Time* is one currency in which digital sign cost can be differentiated. Time translates into information in at least three different ways: aggregation (total time spent in activism), relative expenditure (time spent on the forums as a proportion of time available in a given period), and skill acquisition (time spent learning). From this I propose a tentative taxonomy of time-related virtual signs, consisting of historical signs, presence signs, and knowledge signs. Historical signs are things that show a person has been in the game for a long time. Presence signs show that a person is able to maintain a continuous or at least frequent presence on the forums over a certain period. Knowledge signs are displays of hard-to-earn knowledge. Future research may help evaluate this hypothetical categorization.

To be sure, this study has only scratched the surface of the topic of trust among high-risk activists online. More research is needed, especially on trust on other digital platforms and among other types of political activists online. It would be interesting to see, for example, whether low-bandwidth media such as Twitter see less interpersonal trust between jihadis than higher-bandwidth media such as Paltalk. Another question is whether platforms with different interface constraints (such as the ability, on Facebook, to see other users' 'friends') produce different trust dynamics or distrust mitigation strategies. Similarly, by studying forums for other types of militants, we might be able to observe how trust dynamics play out under different levels of repression, or to what extent ideological–cultural specificities affect signalling knowledge. So far, existing studies on non-jihadi forums such as the far-right *Stormfront* (Bowman-Grieve 2009) and dissident Irish republican websites (Bowman-Grieve and Conway 2012) do report symptoms of trust problems, but not on the scale documented in this chapter.

The decline of jihadi forums did not put an end to effective jihadi communication in the online sphere. In the early 2010s, the emergence of social media ushered in a 'digital empowerment revolution' for non-state actors (Hegghammer 2016). A combination of factors—including the rapid proliferation of new platforms, legal constraints on intelligence operations against Silicon Valley companies, and the general increase in jihadi activity due to the Syria war—limited governments' ability to police jihadi activity online. The overall level of repression was particularly low between 2012 and 2015; in this period, jihadis operated much more freely online than they had been able to just a few years previously. The effect of this development was most obvious on the recruitment of foreign fighters to Syria (Aran 2014; Abu al-Khayr 2013). Social media notably allowed early travellers to communicate directly with prospective recruits through a variety of digital channels. They chatted on Facebook with friends back home (Cachia et al. 2014), tweeted about life at the front line (Klausen 2015), ran Tumblr blogs with travel advice (Sherlock 2013), and took questions on Ask.fm on how to get to Syria (Paraszczuk 2014). This 'bridgehead capability' was likely one of the reasons why the Syria war attracted record numbers of Islamist foreign fighters. However, the tide would turn again in the mid-2010s as states sought to recapture the lost online terrain. From around 2015, Western governments significantly ramped up their digital counterefforts and enacted legislation to bring technology companies on board (Jones and Solomon 2014; Moutot 2015; Hegghammer 2021). By 2018, online jihadi activity was significantly reduced, and paranoia and distrust had returned. The situation remains largely the same as this book goes to press in 2023.

So: is the Internet a blessing or a curse for rebels? The answer seems to be that it depends. The digital domain has certain characteristics which complicate trust-building, but the acuteness of the problem varies according to the level of digital policing and the availability of stealth technologies. If history is a guide, rebel Internet exploitation will continue to fluctuate between periods of empowerment and periods of weakness as governments and their opponents remain locked in a cat-and-mouse technology race.

References

Abu al -Khayr, Waleed. 2013. 'Al-Qaeda in Syria Recruiting Foreign Fighters Online'. *Al-Shorfa.com*, December.

Abu Jarir. 2009. *al–mu'arrifat al–masbuha fi'l–muntadayat wa'l–shabakat al–jihadiyya* [Fake Profiles on Jihadi Networks and Forums]. Ansar Al-Mujahidin, 11 August. http://www.as-ansar.com/vb/showthread.php?p=22750#post22750.

AIVD. 2012. *Jihadism on the Web: A Breeding Ground for Jihad in the Modern Age.* The Hague: The Hague General Intelligence and Security Service.

al-Jihad, Sayf. 2010. '*al–ta'amul ma al–jawasis* [Dealing with Spies]'. *Al-Falluja.* 26 January. http://202.71.102.68/alfaloj/vb/showthread.php?t=100961.

Al-Libi, Abu Yahya. 2009. '*al–mu'allim fi hukm al–jasus al–muslim* [Pointer Regarding the Ruling on the Muslim Spy]'. *Al-Fajr Media.* https://tawhed.ws/dl?i=y34deswq.

al-Maqdisi, Abu al-Nur. 2009. *hukm al–jasus fi'l–islam* [Ruling on the Spy in Islam]. https://archive.org/details/ar001.

Amble, John Curtis. 2012. 'Combating Terrorism in the New Media Environment'. *Studies in Conflict & Terrorism* 35 (5): 339–353.

Anderson, D. Scott. 2006. 'What Trust Is in These Times? Examining the Foundations of Online Trust'. *Emory Law Review* (54): 1441–1474.

Anonymous. 2010. '10 Methods To Detect And Foil The Plots Of Spies'. *Ansar Al-Mujahideen.* http://ansar1.info/showthread.php?t=29715.

Aran, Isha. 2014. 'Becoming Mujahida: Recruiting Western Wives for Islamist Jihadists'. *Jezebel.com.* 8 August.

Arquilla, John and David Ronfeldt (eds). 2001. *Networks and Netwars: The Future of Terror, Crime, and Militancy.* Santa Monica, CA: RAND Corporation.

Awan, Akil, Andrew Hoskins, and Ben O'Loughlin. 2011. *Radicalisation and Media: Connectivity and Terrorism in the New Media Ecology.* London: Routledge.

Bacharach, Michael and Diego Gambetta. 2003. 'Trust in Signs'. In *Trust in Society*, edited by Karen S. Cook, 148–184. New York: Russell Sage Foundation Publications.

Bailenson, Jeremy N. and Andrew Beall. 2006. 'Transformed Social Interaction: Exploring the Digital Plasticity of Avatars'. In *Avatars at Work and Play: Collaboration and Interaction in Shared Virtual Environments*, edited by Ralph Schroeder and Ann-Sofie Axelsson, 1–16. Dordrecht: Springer.

Batya, Friedman and Peter H. Kahn Jr. 2000. 'Trust Online'. *Communications of the ACM* 43 (12): 34–40.

Bergin, Anthony, Sulastri Bte Osman, Carl Ungerer, and Nur Azlin Muhammed Yasin. 2009. 'Countering Internet Radicalisation in Southeast Asia'. *Special Report* 22. Canberra: Australian Strategic Policy Institute.

Biyokulule. 2009. 'Forum Member Warns Pal Talk Program Is "Spying" Tool Against Mujahidin'. *Biyokulule.com.* 3 April.

Borum, Randy. 2010. 'The Science of Interpersonal Trust'. *Mental Health Law & Policy Faculty Publications.* http://scholarcommons.usf.edu/mhlp_facpub/574.

Bos, Nathan, Judy Olson, Darren Gergle, Gary Olson, and Zach Wright. 2002. 'Effects of Four Computer-Mediated Communications Channels on Trust Development'. In *Proceedings of Conference on Human Factors in Computer Systems* 4 (1): 135–140. Minneapolis, MN: ACM Press.

Bowman-Grieve, Lorraine. 2009. 'Exploring "Stormfront": A Virtual Community of the Radical Right'. *Studies in Conflict & Terrorism* 32 (11): 989–1007.

Bowman-Grieve, Lorraine and Maura Conway. 2012. 'Exploring the Form and Function of Dissident Irish Republican Online Discourses'. *Media, War & Conflict* 5 (1): 71–85.

Brachman, Jarret. 2006. 'High-Tech Terror: Al-Qaeda's Use of New Technology'. *Fletcher Forum of World Affairs* 30 (2): 149–164.

Bunt, Gary R. 2003. *Islam in the Digital Age: E-Jihad, Online Fatwas and Cyber Islamic Environments*. London: Pluto Press.

Bunt, Gary R. 2009. *iMuslims: Rewiring the House of Islam*. Chapel Hill, NC: University of North Carolina Press.

Cachia, Rebecca, Susan Clandillon, Inte Gloerich, Cristel Kolopaking, Indre Lauciute, and Rose Rowson. 2014. 'Facebook Pages in Europe: A Tool for the Jihad in Syria'. *Digital Methods Initiative*. https://wiki.digitalmethods.net/Dmi/Winter2014Project12.

Castells, Manuel. 2009. *The Power of Identity: The Information Age: Economy, Society, and Culture*, Volume II. 2nd edn. Malden, MA: Wiley-Blackwell.

Chen, Hsinchun, Dorothy Denning, Nancy Roberts, Catherine A. Larson, Ximing Yu, and Chun-Neng Huang. 2011. 'The Dark Web Forum Portal: From multilingual to video'. *Proceedings of 2011 IEEE International Conference on Intelligence and Security Informatics, Beijing, China, 2011*: 7–14.

Conway, Maura. 2006. 'Terrorist Use of the Internet and Fighting Back'. *Information and Security* 19: 9–30.

Cornwell, B. and D. C. Lundgren. 2001. 'Love on the Internet: Involvement and Misrepresentation in Romantic Relationships in Cyberspace vs. Realspace'. *Computers in Human Behavior* 17 (2): 197–211.

Corritore, Cynthia L., Beverly Kracher, and Susan Wiedenbeck. 2003. 'On-Line Trust: Concepts, Evolving Themes, a Model'. *International Journal of Human-Computer Studies* 58 (6): 737–758.

Cruickshank, Paul and Tim Lister. 2013. 'Al Qaeda Calling?' *CNN.com*, 8 August.

Denning, Dorothy. 2010. 'Terror's Web: How the Internet Is Transforming Terrorism'. In *Handbook of Internet Crime*, edited by Yvonne Jewkes and Majid Yar, 194–213. Cullompton: Willan.

Diani, Mario. 2000. 'Social Movement Networks Virtual and Real'. *Information, Communication & Society* 3 (3): 386–401.

Donath, Judith. 1998. 'Identity and Deception in the Virtual Community'. In *Communities in Cyberspace*, edited by Peter Kollock and Marc Smith, 37–68. London: Routledge.

Drennan, Shane and Andrew Black. 2007. 'Jihad Online: The Changing Role of the Internet'. *Jane's Intelligence Review* 18 (8): 16–20.

Feng, Jinjuan, Jonathan Lazar, and Jenny Preece. 2004. 'Empathy and Online Interpersonal Trust: A Fragile Relationship'. *Behaviour & Information Technology* 23 (2): 97–106.

Finsnes, Cecile. 2010. 'What Is Audio-Visual Jihadi Propaganda? An Overview of the Content of FFI's Jihadi Video Database'. *FFI Report* 2010/00960. Kjeller: FFI.

Fox, Susannah. 2000. 'Trust and Privacy Online'. *Pew Research Center: Internet, Science & Tech*, 20 August.

Frankel, Tamar. 2001. 'Trusting and Non-Trusting on the Internet'. *Boston University Law Review* 82 (2): 457.

Frankel, Tamar. 2009. 'Trust and the Internet'. In *ABA Guide to International Business Negotiations*, edited by James R. Silkenat, Jeffrey Aresty, and Jacqueline Klosek, 7–31. Chicago, IL: ABA Book Publishing.

Galanxhi, Holtjona and Fiona Fui-Hoon Nah. 2007. 'Deception in Cyberspace: A Comparison of Text-Only vs. Avatar-Supported Medium'. *International Journal of Human-Computer Studies* 65 (9): 770–783.

Gambetta, Diego. 2005. 'Deceptive Mimicry in Humans'. In *Perspectives on Imitation: From Neuroscience to Social Science*. Volume 2: *Imitation, Human Development and Culture*, edited by Susan L. Hurley and Nick Chater, 221–242. Cambridge, MA: MIT Press.

Gardham, Duncan. 2011. 'MI6 Attacks Al-Qaeda in "Operation Cupcake"'. *Daily Telegraph*, 2 June.

Grazioli, Stefano and Sirkka L. Jarvenpaa. 2000. 'Perils of Internet Fraud: An Empirical Investigation of Deception and Trust with Experienced Internet Consumers'. *IEEE Transactions Investigation of Deception and Trust with Experienced Internet Consumers* 30 (4): 395–410.

Green, Melanie C. 2007. 'Trust and Social Interaction on the Internet'. In *Oxford Handbook of Internet Psychology*, edited by Adam Joinson, Katelyn McKenna, Tom Postmes, and Ulf-Dietrich Reips, 43–52. Oxford: Oxford University Press.

Hancock, Jeffrey T. 2007. 'Digital Deception: Why, When and How People Lie Online'. In *Oxford Handbook of Internet Psychology*, edited by Adam Joinson, Katelyn McKenna, Tom Postmes, and Ulf-Dietrich Reips, 289–302. Oxford: Oxford University Press.

Hazan. 2010. Tension, Suspicion Among Jihadi Websites Following Infiltration, Collapse of Several Sites. *MEMRI Inquiry and Analysis Series* 625, 14 July.

Hegghammer, Thomas. 2013a. 'The Recruiter's Dilemma Signalling and Rebel Recruitment Tactics'. *Journal of Peace Research* 50 (1): 3–16.

Hegghammer, Thomas. 2013b. 'Syria's Foreign Fighters'. *Foreign Policy*, 9 December.

Hegghammer, Thomas. 2016. 'The Jihadist Digital Empowerment Revolution'. In *The Foreign Fighters Phenomenon and Related Security Trends in the Middle East.* World Watch: Expert Notes 01-01-2016. Ottawa: Canadian Security and Intelligence Service. https://www.csis.gc.ca/pblctns/wrldwtch/2016/20160129-en.php.

Hegghammer, Thomas. 2021. 'Resistance Is Futile: The War on Terror Supercharged State Power'. *Foreign Affairs* 100 (5): 44–52.

Hoffman, Bruce. 2006. '*The Use of the Internet By Islamic Extremists*'. Testimony presented to the House Permanent Select Committee on Intelligence. Washington, DC: RAND Corporation.

ICT/JWMG Team. 2012. *In the Depths of Jihadist Web Forums: Understanding a Key Component of the Propaganda of Jihad.* Herzlia: International Institute for Counter-Terrorism.

Jakes, Lara and Adam Goldman. 2013. 'Terrorists Turn to Online Chat Rooms to Evade US'. *Associated Press*, 14 August.

Jones, Sam and Erika Solomon. 2014. 'Isis Closes the Cyber Blackout Blinds to Avoid Attack'. *Financial Times*, 17 October.

Kenney, Michael. 2010. 'Beyond the Internet: Metis, Techne, and the Limitations of Online Artifacts for Islamist Terrorists'. *Terrorism and Political Violence* 22 (2): 177–197.

Kimmage, Daniel. 2008. 'The Al-Qaeda Media Nexus: The Virtual Network behind the Global Message'. Special Report. Washington, DC: Radio Free Europe/Radio Liberty.

Klausen, Jytte. 2015. 'Tweeting the Jihad: Social Media Networks of Western Foreign Fighters in Syria and Iraq'. *Studies in Conflict & Terrorism* 38 (1): 1–22.

Kling, Rob. 1996. 'Social Relationships in Electronic Forums: Hangouts, Salons, Workplaces, and Communities'. In *Computerization and Controversy: Value Conflicts and Social Choices*, edited by Rob Kling, 426–454. 2nd edn. San Diego, CA: Morgan Kaufmann.

Koehn, Daryl. 2003. 'The Nature of and Conditions for Online Trust'. *Journal of Business Ethics* 43 (1–2): 3–19.

Kohlmann, Evan. 2008. 'Al-Qaida's Myspace: Terrorist Recruitment on the Internet'. *CTC Sentinel* 1 (2): 8–9.

Labi, Nadya. 2006. 'Jihad 2.0'. *The Atlantic*, July/August.

Lappin, Yaakov. 2011. *Virtual Caliphate: Exposing the Islamist State on the Internet.* Dulles, VA: Potomac Books.

Lea, Martin and Russell Spears. 1995. 'Love at First Byte? Building Personal Relationships over Computer Networks'. In *Under-Studied Relationships: Off the Beaten*

Track, edited by Julia T. Wood and Steve Duck, 197–233 Thousand Oaks, CA: SAGE Publications.

Lia, Brynjar. 2006. 'Al-Qaeda Online: Understanding Jihadist Internet Infrastructure'. *Jane's Intelligence Review* 18 (1): 14–19.

Lusthaus, Jonathan. 2012. 'Trust in the World of Cybercrime'. *Global Crime* 13 (2): 71–94.

Mantel, Barbara. 2009. 'Terrorism and the Internet: Should Web Sites That Promote Terrorism Be Shut Down?' *CQ Global Researcher* 3 (11): 129–153.

Marx, Gary. 1974. 'Thoughts on a Neglected Category of Social Movement Participant: The Agent Provocateur and the Informant'. *American Journal of Sociology* 80 (2): 402–442.

Marx, Gary. 2013. 'Agents Provocateurs as a Type of Faux Activist'. In *The Wiley-Blackwell Encyclopedia of Social and Political Movements*, edited by David A. Snow, Donatella Della Porta, Bert Klandermans, and Doug McAdam. Malden, MA: Wiley-Blackwell.

Milton, Daniel. 2022. 'Truth and Lies in the Caliphate: The Use of Deception in Islamic State Propaganda'. *Media, War, and Conflict* 15 (2): 221–237.

Misztal, Barbara. 1996. *Trust in Modern Societies: The Search for the Bases of Social Order*. Cambridge, MA: Polity.

Moubayed, Sami. 2006. 'Syria's Abu Al-Qaqa: Authentic Jihadist or Imposter?' *Jamestown Terrorism Focus* 3 (25): 6–8.

Moutot, Michel. 2015. 'Jihadis Increasingly Wary of Internet Breaches, Experts Say'. *Daily Star*, 2 February.

Musawi, Mohammed Ali. 2010. '*Cheering for Osama: How Jihadists Use Internet Discussion Forums*'. London: Quilliam Foundation.

Myers, David. 1987. '"Anonymity Is Part of the Magic": Individual Manipulation of Computer-Mediated Communication Contexts'. *Qualitative Sociology* 10 (3): 251–266.

Nacos, Brigitte L. 2002. *Mass-mediated Terrorism: The Central Role of the Media in Terrorism and Counterterrorism*. Lanham, MD: Rowman & Littlefield Publishers.

Nakashima, Ellen. 2010. 'Dismantling of Saudi-CIA Web Site Illustrates Need for Clearer Cyberwar Policies'. *Washington Post*, 19 March.

Naquin, Charles E. and Gaylen D. Paulson. 2003. 'Online Bargaining and Interpersonal Trust'. *The Journal of Applied Psychology* 88 (1): 113–120.

Nissenbaum, Helen. 2001. 'Securing Trust Online: Wisdom or Oxymoron'. *Boston University Law Review* 81 (3): 635–664.

Paraszczuk, Joanna. 2014. 'The Frequently Asked Questions of Aspiring Jihadists'. *The Atlantic*, 12 November.

Parks, Malcolm R. and Kory Floyd. 1996. 'Making Friends in Cyberspace'. *Journal of Communication* 46 (1): 80–97.

Policy Planners' Network. 2011. '*Radicalisation: The Role of the Internet*'. London: Institute for Strategic Dialogue.

Ranstorp, Magnus. 2004. 'Al-Qaida in Cyberspace: Future Challenges of Terrorism in the Information Age'. In *Terrorism in the Information Age—New Frontiers*, edited by Lars Nicander and Magnus Ranstorp, 58–72. Stockholm: Swedish National Defence College.

Ranstorp, Magnus. 2007. 'The Virtual Sanctuary of Al-Qaida and Terrorism in an Age of Globalization'. In *International Relations and Security in the Digital Age*, edited by Johan Eriksson and Giampiero Giacomello, 31–56. London: Routledge.

Rawnsley, Adam. 2010. 'Taliban Webmaster: We've Been Hacked!' *Wired.com*, 10 June.

Rawnsley, Adam. 2011. '"Spyware" Incident Spooks Jihadi Forums'. *Wired.com*, 1 September.

Rocco, Elena. 1998. 'Trust Breaks Down in Electronic Contexts but Can Be Repaired by Some Initial Face-to-Face Contact'. In *CHI '98: Proceedings of the SIGCHI Conference on Human Factors in Computing Systems*. Los Angeles, CA: ACM Press.

Rogan, Hanna. 2006a. 'Al-Qaeda's Online Media Strategies: From Abu Reuter to Irhabi 007'. *FFI Report* 2007/02729. Kjeller: FFI.

Rogan, Hanna. 2006b. 'Jihadism Online—A Study of How Al-Qaida and Radical Islamist Groups Use the Internet for Terrorist Purposes'. *FFI Report* 2006/00915. Kjeller: FFI.

Rossmiller, Shannen. 2007. 'My Cyber Counter-Jihad'. *Middle East Quarterly* 14 (3): 43–48.

Ryan, Thomas. 2010. 'Getting in Bed with Robin Sage'. *Black Hat Briefings and Training*. https://media.blackhat.com/bh-us-10/whitepapers/Ryan/BlackHat-USA-2010-Ryan-Getting-In-Bed-With-Robin-Sage-v1.0.pdf.

Sageman, Marc. 2004. *Understanding Terror Networks*. Philadelphia, PA: University of Pennsylvania Press.

Schmitt, Eric and Thom Shanker. 2011. *Counterstrike: The Untold Story of America's Secret Campaign Against Al Qaeda*. New York: Times Book.

Seib, Philip. 2008. 'The Al-Qaeda Media Machine'. *Military Review* 88 (3): 74–80.

Seib, Philip and Dana M. Janbek. 2011. *Global Terrorism and New Media: The Post-Al Qaeda Generation*. London: Routledge.

Shahar, Yael. 2007. 'The Internet as a Tool for Intelligence and Counter-Terrorism'. In *Hypermedia Seduction for Terrorist Recruiting: Volume 25, NATO Science for Peace and Security Series: Human and Societal Dynamics*, edited by B. Ganor, K. Von Knop, and C. Duarte, 104–117. Amsterdam: IOS Press.

Sherlock, Ruth. 2013. 'What Every Jihadi in Syria Needs: Hair Gel, an iPad and Kit-Kats'. *Daily Telegraph*, 26 November.

Swann, Steve. 2011. 'Rajib Karim: The Terrorist inside British Airways'. *BBC News Online*, 28 February.

Sztompka, Piotr. 2000. *Trust: A Sociological Theory*. Cambridge and New York: Cambridge University Press.

Talbot, David. 2005. 'Terror's Server: How Radical Islamists Use Internet Fraud to Finance Terrorism and Exploit the Internet for Jihad Propaganda and Recruitment'. *MIT Technology Review*, 27 January.

Tarrow, Sidney. 2005. *The New Transnational Activism*. New York: Cambridge University Press.

Theohary, Catherine A. and John Rollins. 2011. *'Terrorist Use of the Internet: Information Operations in Cyberspace'*. CRS Report for Congress. Washington, DC: Congressional Research Service.

Thomas, Timothy L. 2003. 'Al Qaeda and the Internet: The Danger of "Cyberplanning"'. *Parameters* 23 (1): 112–123.

Torres Soriano, Manuel R. 2012. 'The Vulnerabilities of Online Terrorism'. *Studies in Conflict & Terrorism* 35 (4): 263–277.

Ulph, Stephen. 2004. 'Forum Warnings of a Spy Website'. *Jamestown Terrorism Focus* 1 (9).

Ulph, Stephen. 2006a. 'Fears of Intelligence Penetration of the GIMF'. *Jamestown Terrorism Focus* 3 (16): 5–6.

Ulph, Stephen. 2006b. 'Intelligence War Breaks out on the Jihadi Forums'. *Jamestown Terrorism Focus* 3 (14): 4–5.

Ulph, Stephen. 2006c. 'Mujahideen Explain Away Failures of the Abqaiq Attack'. *Jamestown Terrorism Focus* 3 (9): 3–4.

Van Gelder, Lindsy. 1985. 'The Strange Case of the Electronic Lover'. *Ms. Magazine*, October.

Van Lear, Jeroen and Peter Van Aelst. 2009. 'Cyber-Protest and Civil Society: The Internet and Action Repertoires in Social Movements'. In *Handbook of Internet Crime*, edited by Yvonne Jewkes and Majid Yar, 230–254. Cullompton: Willan.

Vijayan, Jaikumar. 2008. 'Updated Encryption Tool for Al-Qaeda Backers Improves on First Version, Researcher Says'. *Computerworld.com*, 4 February.

Wang, Ye Diana and Henry H. Emurian. 2005. 'An Overview of Online Trust: Concepts, Elements, and Implications'. *Computers in Human Behavior* 21 (1): 105–125.

Warrick, Joby. 2011. *The Triple Agent: The Al-Qaeda Mole Who Infiltrated the CIA*. New York: Doubleday.

Weimann, Gabriel. 2006. *Terror on the Internet: The New Arena, the New Challenges*. Washington, DC: United States Institute of Peace Press.

Weimann, Gabriel. 2007. 'Using the Internet for Terrorist Recruitment and Mobilization'. In *Hypermedia Seduction for Terrorist Recruiting:* Volume 25, *NATO Science for Peace and Security Series: Human and Societal Dynamics*, edited by B. Ganor, K. Von Knop, and C. Duarte, 47–58. Amsterdam: IOS Press.

Weisburd, Aaron. 2006a. 'Irhabi007 as a Case-Study in Jihadi Counter-Intelligence Efforts'. *Internet-Haganah.* 11 March.

Weisburd, Aaron. 2006b. 'Poor 007'. *Internet-Haganah.* 11 March.

Weisburd, Aaron. 2007. 'The Shifting Sands of the Global Jihad Online'. In *Hypermedia Seduction for Terrorist Recruiting:* Volume 25, *NATO Science for Peace and Security Series: Human and Societal Dynamics*, edited by B. Ganor, K. Von Knop, and C. Duarte, 154–165. Amsterdam: IOS Press.

Williams, Molly. 2001. 'A Living Dream'. *Wall Street Journal,* 24 September.

Zanini, Michelle and Sean J. A. Edwards. 2001. 'The Networking of Terrorism in the Information Age'. In *Networks and Netwars: The Future of Terror, Crime, and Militancy*, edited by John Arquilla and David Ronfeldt, 29–60. Santa Monica, CA: RAND Corporation.

Zelin, Aaron Y. 2013. *The State of Global Jihad Online: A Qualitative, Quantitative, and Cross-Lingual Analysis.* Washington, DC: New America Foundation.

4

The Code Word Conundrum in the Northern Ireland Conflict

Heather Hamill

In March and April 1969 two explosive devices were detonated near Belfast, Northern Ireland; one at an electricity substation and another at a reservoir.[1] The press and the British and Irish security services blamed the Irish Republican Army (IRA) for these actions. Then, on 21 October 1969, Thomas McDowell, a member of a Loyalist paramilitary group called the Ulster Protestant Volunteers (UPV) was killed by a bomb he was attempting to plant at a power station near Ballyshannon in Donegal. The device was of the same type as those used to attack the electricity substation and the reservoir. The UPV had designed their actions to make people believe it was the IRA that had planted all the devices. Their aim was to discredit the IRA and the tentative reforms that were being made towards Catholics by Northern Ireland's prime minister, Terence O'Neill. This incident is the first clear example, in the Northern Ireland conflict, of one armed group mimicking another armed group of the opposite religion. These events occurred at the beginning of the conflict in Northern Ireland that became known as 'the Troubles'.

According to Diego Gambetta (2005), deceptive mimicry occurs when a mimic passes him or herself off as being someone they are not. The mimic does this by signalling to others the identity and attributes of the individual being mimicked. In the first part of this chapter, I discuss how, for the most part, paramilitaries were able to use camouflage defensively and aggressively. I then examine a series of instances of *defensive* and *aggressive mimicry*, at both the individual and group level. In the second part of the chapter, I focus on code words, which were used by paramilitaries in their communications

---

[1] I would like to thank Andy Richards, Ian Cropper, and the *Birmingham Mail* for giving me access to the log book; all those who agreed to be interviewed and Eugene McConville, Jimmy Quinn, Anja Shortland, and Peter Taylor for their advice and facilitation of those interviews. I received very helpful comments from Edmund Chattoe-Brown, Steve Fisher, my fellow contributors to this volume, and especially Diego Gambetta.

Heather Hamill, *The Code Word Conundrum in the Northern Ireland Conflict*. In: *Fight, Flight, Mimic*. Edited by: Diego Gambetta and Thomas Hegghammer, Oxford University Press. © Heather Hamill (2024). DOI: 10.1093/9780191802454.003.0004

with the security forces and the wider public. Code words are included in messages fore-warning of bomb explosions in commercial centres; claims or denials of responsibility for violent actions; and when making statements about strategic decisions within a paramilitary organization. I explore code words as an identity signal and its use as an anti-mimicry device.

Background to the Troubles

Northern Ireland has been the scene of the worst political violence in Western Europe since the Second World War. The death toll between 1969 and the present day is 3,386.[2] The conflict in Northern Ireland has been long-running and of low intensity relative to many other civil wars. The casualty figures are large in aggregate, but 'the Troubles' have lasted over forty years. The struggle has been between Republicans (Catholic militants) and Loyalists (Protestant militants); Republicans and the security forces (British Army and the police) and Loyalists and the security forces. Throughout the conflict, the IRA was the largest and most active paramilitary group within Republicanism. Following the IRA's first ceasefire in 1994, two dissident IRA splinter groups were formed: the Continuity IRA (CIRA) and the Real IRA (rIRA) which split again to form Óglaigh na hÉireann. These dissident groups aim to continue the armed struggle.

On the Protestant side, there are two main pro-state Loyalist paramilitary groups. The larger of the two, the Ulster Defence Association (UDA), also used the cover name of Ulster Freedom Fighters (UFF) to claim responsibility for killing Catholics.[3] The Ulster Volunteer Force (UVF) is smaller in number than the UDA but is much deadlier.[4] Loyalist attacks are mostly sectarian against Catholic civilians and Republican paramilitaries. Republicans will attack Loyalist paramilitaries and political, economic, and military targets. Active service members of the armed groups are overwhelmingly white, male, working class and aged between eighteen and forty.

The other actors are the state security services. The police—originally the Royal Ulster Constabulary (RUC)—is now called the Police Service of Northern Ireland (PSNI). The British Army was deployed on active service in the streets of Northern Ireland in August 1969. The Ulster Defence Regiment (UDR) was an infantry regiment of the British Army which recruited locally from the Protestant population.

---

[2] PSNI, 2015. www.psni.police.uk/index/updates/updates_statistics/updates_security_situation_stat istics.htm (accessed 25 March 2015).
[3] CAIN, 2006. http://cain.ulst.ac.uk/othelem/organ/uorgan.htm#uda (accessed 15 January 2007).
[4] CAIN, 2015. http://cain.ulst.ac.uk/othelem/organ/uorgan.htm#uda (accessed 15 January 2007).

Camouflage

The religious segregation in Northern Ireland is well documented (Hughes et al. 2007). Thirty-five to forty per cent of the Northern Irish population live in completely segregated neighbourhoods (Hewstone et al. 2004, 265–292). Even in the current post-conflict phase there are forty-nine 'peace walls' in Belfast demarcating Catholic and Protestant neighbourhoods (Rutherford and Bowyer 2011). Religious affiliation is self-evident by virtue of living in the area and painted wall murals demarcate an organization's presence and residents' loyalties. But identifying a paramilitary from an 'ordinary citizen'— the 'identification problem' (Kalyvas 2006, 89–91)—is a significant challenge. Going unnoticed within their communities is effortless for paramilitaries. They are simultaneously a working-class Catholic or Protestant and a member of an armed group. They do not need to modify their appearance or behaviour to blend into the background. Membership of the IRA peaked at around 1500 in the mid-1970s. In 1994, at the time of the ceasefire, membership was approximately 500 with a smaller number being 'active' members.[5] Most of these members were inconspicuous, moving freely within their community and appearing to be nothing more than a working-class Catholic. Active service units operating in Britain and mainland Europe also adopted this strategy. Likewise, most of the 30,000 estimated members of the UDA and several hundred members of the UVF[6] remained camouflaged in their communities. The aim of this camouflage was defensive. They wanted to avoid being identified and targeted for surveillance or attack by the security forces or by a paramilitary group of the opposite religion.

Camouflage also enabled logistical actions to be carried out without interruption. Paramilitaries moved weapons in cars with their wives and children in the passenger seats. They were simultaneously an ordinary family on their way to visit their grandparents and moving weapons from one safe house to another.

While camouflage protects the paramilitary, its use puts ordinary people at risk of being targeted indiscriminately. The British Army was unable to distinguish paramilitaries from the 'ordinary' population and this led to internment: the arrest and detention without trial of people suspected of being members of illegal paramilitary groups between 1971 and 1975. During this period, a total of 1981 people were detained; 1874 were Catholic/Republican; only 107 were Protestant/Loyalist.[7] In 1978, the European Court of Human

---

[5] CAIN, 2015. http://cain.ulst.ac.uk/issues/violence/paramilitary2.htm (accessed 18 February 2015).
[6] CAIN, 2015. http://cain.ulst.ac.uk/issues/violence/paramilitary2.htm (accessed 18 February 2015).
[7] CAIN, 2015. http://www.cain.ulst.ac.uk/events/intern/sum.htm (accessed 23 March 2015).

Rights ruled that the use of rough interrogation techniques during intern-ment in Northern Ireland constituted 'inhuman and degrading punishment' (Weitzer 1995, 178). The British Army relied on lists of 'suspects' drawn up by the RUC Special Branch, but their information was of extremely poor qual-ity. Coogan (1995, 126) describes how 'The army quite often simply picked up the wrong people, a son for a father, the wrong "man with a beard living at No. 47" and so on'.

Internment—also known as Operation Demetrius—was disproportion-ately directed towards the IRA. Undoubtedly, this reflected the government's punitive bias towards Republicans as opposed to pro-state Loyalists. More-over, the Protestant police force shared a religious, social, cultural, and even familial background with Loyalists and so had better information as to who was a Loyalist paramilitary. The Army had less need to apply a drag-net strategy to this side of the conflict. This differential treatment carried on throughout the Troubles and further polarized the two communities. Working-class Catholics were subjected to house raids, the use of stop-and-search powers, and detention without trial inscribed in various statutory powers enacted between 1973 and the present day (Campbell and Connolly 2006, 945–946).

The immediate effect of internment was an increase in the amount of vio-lence. Interment began in August 1971; the number of explosions increased from 79 in July 1971 to 142, 186, 155, 117, and 123 in August to Decem-ber 1971 respectively (Spjut 1986, 716). It radicalized Catholics, reduced the number of Catholic moderates, cemented a deep-rooted fear and sus-picion of the police that endures today (Hamill 2011), and boosted IRA recruitment.

Paramilitaries most often used camouflage defensively, but they also used their ability to blend in, to target others. The police recorded 19,187 bombs and incendiary devices (excluding devices that were neutralized) having been used in Northern Ireland between 1969 and the time of writing[8] (excluding devices utilized in the rest of the UK and Europe). Most of these devices were placed in public places. To plant a bomb, the bomber needs to pass as a relaxed ordinary citizen. When a former IRA member planted an incendiary device in a large furniture shop he '... just walked into a shop, set the bag [containing the device] down, and walked out'.[9] Unlike in the Israeli–Palestinian conflict (Benmelech and Berrebi 2007), bombers were not caught in the act because of a failure to signal normality. Prior intelligence

---

[8] http://www.psni.police.uk/index/updates/updates_statistics/updates_security_situation_statistics. htm (accessed 23 March 2015).
[9] Interview with former IRA member, Belfast, 22 August 2012.

and technical incompetence, when bombs exploded prematurely, led to their detection.

Bombers exploited their shared physical and behavioural traits with the opposite religion and 'ordinary' British citizens. This strategy was exemplified in 1984 when Patrick Magee, a member of the IRA, planted a bomb in the Grand Hotel in Brighton. The target was Prime Minister Margaret Thatcher and her cabinet who were attending the Conservative Party Conference. The explosion killed five people, but Thatcher escaped unharmed. Magee had excellent camouflage. Born in Belfast, he moved to England aged four, so he had a genuine English accent (Hatterstone 2001). He used a false name of Roy Walsh when he checked into the hotel. This character was made up; he was not impersonating anyone. He also checked into the hotel twenty-four days prior to the conference and the bomb, planted under the bath in his room, had a long delay timer. By placing the bomb so far in advance, he further avoided suspicion and the intense screening put in place closer to the conference date.

Defensive Mimicry

Catholic civilians were most at risk of being murdered throughout the Troubles, and the greatest proportion of civilian murders were committed in North Belfast. This part of Belfast is a patchwork of Catholic and Protestant neighbourhoods and has more sectarian interfaces than anywhere else in the city (Sluka 2000, 136). If you happened to stray into the 'wrong' street, you were at risk from paramilitaries on the prowl looking for victims. Despite people being killed in these circumstances, there are few examples of defensive mimicry when a member of one religion mimics a member of another religion to avoid harm. There may, of course, have been unsuccessful attempts but the mimic did not live to tell the tale. As in Rwanda (see Laitin in this volume), strong group boundaries meant that there was high social knowledge of who belonged to what religion and where they fitted into the social order. In June 1973 Daniel O'Neill, a Catholic, was on his way home after a night out when he was picked up by Loyalists in a staunchly Protestant neighbourhood. He was well known in both communities because of his job as a television-rent collector (McKittrick et al. 2007, 366). His assassins knew he was a Catholic so mimicry would have been futile. As the number of victims increased, residents avoided risk by assiduously staying in their neighbourhoods. In 1974, *Republican News* published a seven-point list of precautions that Catholics should take including 'Never go out at night alone in areas recognized to be the assassin's murder ground' (quoted in Sluka 2000, 138). Even if a victim was cool enough to be able to display the correct signs under pressure the local knowledge of who is who raised the probability of being discovered. John McIver was a Scottish Catholic who for years hid his religion from his Protestant friends by, among other things, being a staunch supporter of Glasgow Rangers Football Club.[10] Four months before his murder by Loyalists in 1992 he cried as he told a friend: 'They have found out I'm a Catholic' (*Belfast Telegraph* 2001).

Prison Escapes

Incarcerated Republicans viewed themselves as prisoners of war (POWs) and thus, in company with all POWs, had a duty to escape. IRA prisoners escaped successfully on forty-five occasions between 1971 and 1997 (Kelly 2013, 319–322). They tunnelled out of prison,[11] climbed over perimeter walls using homemade ropes fashioned from bed sheets, were lifted out by helicopter, blasted their way out, and overpowered guards with firearms smuggled in from outside.[12] They also used defensive mimicry in prison escapes. In 1973, John Francis Green escaped dressed as his brother, a priest. His brother Father Gerrard Green, who visited him prior to the escape, was found tied up in the prison (Coogan 2000, 405). Brendan Hughes escaped in 1973. In order to buy himself some time before the authorities realized he had escaped, he had another inmate mimic him, or act as his double (see Gambetta and Hegghammer, Chapter 1 this volume): 'I took my clothes off, gave them to this guy who had the same big moustache as me and black hair', he said (Moloney 2010, 159). In 1997, Liam Averill escaped from the Maze Prison dressed as a woman. He escaped after a Christmas party for prisoners' children. He took advantage of the prison authorities' belief in a 'tacit agreement with all the factions in the jail that the privilege of parties would not be abused' (House of Commons 1998, 12) and so perhaps the guards were less vigilant than normal. Averill wore a wig, make-up, and woman's clothes and five prison guards failed to spot his disguise (AnPhoblacht 1997). Escapees have also worn uniforms to mimic British soldiers and prison officers (McEvoy 2001, 54). Thirty-eight prisoners escaped from the Maze Prison

---

[10] Protestants support Glasgow Rangers Football Club and Catholics support Glasgow Celtic Football Club.

[11] In 1983, it was reported that there were 200 tunnels underneath Long Kesh Prison (also known as the Maze Prison), so many that the place almost collapsed (Coogan 2000, 529).

[12] Unlike Republicans, Loyalist prisoners have made very few escape attempts. For a fuller discussion of the reasons why see McEvoy (2001, 64–67).

in one break-out in 1983. The escapees and their rearguard support wore prison officers' uniforms but when fighting broke out between them and the genuine prison officers, five prison officers were stabbed and two were shot. In the confusion, the British soldiers in the watchtowers overlooking the prison could not distinguish between the genuine prison officers and the mimics (Kelly 2013, 127).

Aggressive Individual Mimicry

Aggressive mimicry is when an individual mimics another individual to harm someone else. McKittrick's (McKittrick et al. 2007) anthology of deaths in Northern Ireland describes in detail the circumstances of 3521 deaths that have resulted from the conflict between 1966 and 2006. This work contains information on the perpetrator(s), what organization they belonged to, and details as to how the murder was carried out. I coded each murder as to what organization was responsible, the status and religion of the victim, whether mimicry occurred, and what form it took. Only instances where the killer assumes another identity are counted as aggressive mimicry. For analytical clarity, bombers who did not assume another's identity or modify their appearance and killers wearing masks to hide their identity, or camouflaged as ordinary citizens are not counted in this analysis. According to this data, aggressive mimicry occurred in 124 cases or four per cent of the total number of deaths. The victims were all shot. Thirty-four per cent of these instances of aggressive mimicry occurred between 1973 and 1976, when the overall numbers of deaths also peaked. The mimics were members of the IRA (forty-four per cent), the UVF (twenty-nine per cent) and the UDA (seventeen per cent). Sixty-nine per cent of the victims of this kind of aggressive mimicry were civilians and of those sixty-three per cent were Catholic and twenty-eight per cent were Protestant. The UVF carried out forty-nine per cent of these killings of Catholic civilians, and the UDA perpetrated twenty-six per cent. We can conclude therefore that aggressive mimicry is most often used in the assassination of Catholic civilians by Loyalist paramilitaries. As Table 4.1 shows, mimics have actively posed as genuine customers seeking a taxi, service in a shop, or a delivery of fast food. Mimics have also passed as postmen, builders, hospital workers, and members of the police and army.

According to this data, the overall number of instances of aggressive mimicry is small relative to the overall number of violent events. It is likely that this kind of mimicry was more prevalent, but the use of mimicry in unsuccessful assassinations has not been recorded.

Table 4.1 Assassinations 1966–2006: Models, Mimics, and Victims (N = 124)

Model	Mimic	Victim
Genuine customer (taxi, fast food, in a shop)	IRA	Civilian/British Army
	UVF	Civilian
	UDA	Civilian /British Army
	RHC	Civilian
	IPLO	Civilian
Taxi driver	UDA	Civilian
	IPLO	Civilian
British Army	UVF	Civilian
UDR	IRA	Civilian /British Army
Council workers	IRA	Civilian
	UVF	Civilian
	UDA	Civilian
Postman	IRA	British Army/UDR
Butchers	IRA	Police
Journalist	Police*	Civilian
Friend of the victim	IRA	Civilian /British Army
Burglary victim	IRA	Police
	IRA	Police
Hitchhiker	RHD	Civilian
Shopkeeper	IRA	British Army
Police	UVF	IRA
IRA	UVF	Civilian
	UDA	Civilian
Hospital workers (wore lab coats)	UDA	Civilian
Students	IRA	Civilian
Paramedics	IRA	British Army
Cyclists	IRA	British Army
Friendly civilian asking for directions	UDA	Civilian
Gardai (Republic of Ireland police)	INLA	INLA
Arabs	IRA	Civilian
Builders	IRA	Civilian
Stranded motorist	UVF	Civilian
Fellow drinker	IRA	Civilian
Drunk	UVF	Civilian
False friend	UVF	Civilian
	UDA	UDA
	IRA	Civilian
'Good time' girls	IRA	British Army
Friend of family member	IRA	British Army
Wedding guest	UVF	Civilian

*The killer was acting alone; this was not an official police operation.
Source: created by author

Successful assassinations require that the killers get within shooting range of their victim, the victim does not sense the threat and escape or retaliate, and the assassins get away unharmed. As has been noted, Catholic civilians were most at risk of being murdered throughout the Troubles. In North Belfast, Loyalist assassins did not have to travel far to kill Catholics as they often lived just a street away from their victims. They could be back in their territory within minutes (Sluka 2000, 136). In these cases, the assassins did not need to use mimicry to affect their kill or facilitate their escape.

Fearing an attack at any time, non-civilians—the police, army personnel, and paramilitaries—took precautions. They varied their patterns of travel and were cautious about who they associated with and distrusted everyone they did not know. They also carried weapons to defend themselves. If they were suspicious, they could shoot first and ask questions later. The assassin's best strategy against these targets was to act as quickly as possible to surprise their victim and not give them any time to react and defend themselves. Drive-by shootings and ambushes worked best, and these did not require any mimicry.

Assassins used aggressive mimicry under any of the following four circumstances. The first was when assassins needed to manoeuvre themselves or the victim into a better position to kill them. Catholic taxi drivers in Belfast were especially vulnerable to acts of aggressive mimicry in this circumstance. Loyalists were willing to take the risk of going into Catholic areas to mimic ordinary Catholic passengers and hail a cab. Once inside the car, they could kill the driver with relative ease (Gambetta and Hamill 2005). Second, mimicry was used when killers needed to persuade their victim to accompany them to another location. Anne-Marie Smyth was a Catholic on her first trip to Belfast. She strayed into a Loyalist pub and, unaware of the dangerous territory she was in, she was duped by other women into believing they, too, were Catholics. They persuaded Anne-Marie to go with them to a house where she was strangled (McKittrick 1997). Third, aggressive mimicry is used when the assassins needed an extra element of surprise to kill their victim. In July 1988, three UVF men dressed as members of the police shot and killed Brendan Davidson in his home. Davidson—an experienced member of the IRA—had served a prison sentence for attempted murder and had escaped an assassination attempt by Loyalists in the previous year (Hearst 1988). If the UVF had appeared at his door in their usual operational garb—balaclavas and boiler suits—he would instantly have known what was happening and would have reacted aggressively. Instead, Davidson would have expected the police to arrest him, not kill him, and so did not immediately act in self-defence. Dressing as the police was also an act of defensive mimicry as it increased the UVF's chances of a safe getaway. Local people would be less hostile to the police than they would be towards suspected Loyalist killers.

Finally, assassins use aggressive mimicry when there is an additional pay-off to the murder. In June 1990, the UFF assassinated Sean Keenan, a Sinn Féin councillor. Keenan lived in a Republican enclave deep within Catholic West Belfast. The hit squad disguised themselves as British soldiers on foot patrol. They dressed in full combats and camouflage and walked through the heart of West Belfast unchallenged (Adair 2007, 57). Army foot patrols were a daily occurrence in Republican areas, so residents did not suspect that there was a different and deadlier enemy in their midst. But the costs of this act of mimicry were very high. Republicans invested a considerable amount of time and effort familiarizing themselves with the movements and practices of the British Army. The UFF needed to source the correct type of uniform and weaponry, and walk in a formation akin to an Army foot patrol. The mimicry worked because Republicans never expected that Loyalists would risk such exposure to attack by trying to pass off as members of the British Army and is another example of mimicry being simultaneously aggressive and defensive. Republicans considered Loyalists an easier target than the British Army. By mimicking the British Army, the UFF reduced the likelihood of being attacked on their retreat. Why were the UFF willing to carry out such a high-risk operation when there were less costly ways to murder Keenan? There was considerable jockeying for power within the UFF at this time (Lister and Jordan 2003). Venturing into Republican territory to kill a member of Sinn Féin in this audacious way showed individual bravery and daring and gained status within the group. The additional status gained from taking part in this operation was the pay-off that made the costs of mimicry worth paying.

Intelligence Gathering

Another reason to adopt a strategy of aggressive mimicry was to gather intelligence. The renegade Loyalist paramilitary, Johnny Adair reported that he made many successful reconnaissance trips into the Catholic enclave in West Belfast. On more than one occasion, he commandeered a black taxi and drove around the area with some women and children to look like an authentic Catholic taxi driver (Adair 2007, 66). On other occasions, he wore a Celtic football club top to pass as a Catholic (Adair 2007, 67). These acts of mimicry were primarily aggressive, but they were also defensive because if he had been spotted, he would have been killed. Adair's added defensive strategy was to avoid talking to any Catholics who might probe him for local knowledge that he may have been unable to provide convincingly.

Members of the intelligence services also mimicked ordinary citizens during surveillance operations. As in Adair's case, their acts of mimicry were primarily aggressive but were also defensive. Captain Robert Nairac was a British soldier involved in intelligence gathering on the IRA in South Armagh in the 1970s. Nairac's method was to pose as Danny McErlaine (or a similar surname), a member of the North Belfast INLA. He assumed that, in the 1970s, the distance of forty miles from South Armagh and North Belfast would impede good information on members of rival Republican paramilitary organizations from flowing smoothly. Although he was English, his Catholic education would help him deflect some probing. Nairac mingled with locals in bars and sang Republican rebel songs to pass off as Danny. However, in May 1977, while undertaking such an operation in the Three Steps Inn near Forkhill, Co. Armagh, the IRA abducted and killed him. According to some reports, IRA sympathizers in the pub became suspicious of him, alerting the IRA because he repeatedly visited the pub's toilet. As he was being abducted the IRA found his Browning pistol, the signature weapon of the British Army, and his cover was blown (*Belfast Telegraph* 2007; McKittrick et al. 2007, 722–724).

The British Secret Intelligence Service (SIS)—also known as MI5—took action to reduce the risk to the individual intelligence officers and make the defensive side of the act of mimicry more robust. For example, they established a fake holiday firm and told a number of Republicans that they had won a free holiday to Spain (Urban 1992, 105). While on these holidays, away from their cadres of support and weaponry, the duped Republicans were much less of a threat. The intelligence officers could reveal their true identity and try to recruit them as agents in relative safety.

Aggressive Group Mimicry

Their similarities in age, gender, appearance and mode of operation presented the paramilitary groups with opportunities to mimic one another. There are instances of paramilitary groups mimicking groups of the opposite religion, coreligionist rival groups, and members of the security forces. These acts of mimicry were carried out to assassinate civilians and paramilitaries of the opposite religion, to discredit other paramilitary groups of the same and opposite religion, and to avoid blame for acts of violence. Paramilitary groups also operated under false flags whereby they used cover groups to avoid taking responsibility for the act of violence.

Mimicking Another Paramilitary Group to Carry out an Assassination

Mimicking an enemy paramilitary group in their territory is an audacious act. Johnny Adair records how the IRA mimicked the UVF to gain access to a house from which to launch an assassination attack on him (Adair 2007, 130). The mimics required courage simply to be in Adair's neighbourhood, but they did not need to do much to dupe the homeowner, who was an almost blind pensioner. Adair also describes how the UDA gained entry to a house in Catholic West Belfast to attack a Catholic by claiming to be the IRA (Adair 2007, 141). The mimics' success depended on the assumption that, to minimize the risk of discovery, paramilitaries would want to spend as little time as possible in enemy territory. When men wearing balaclavas and carrying bags containing weapons asked to use their house, the homeowners assumed they were their 'own' paramilitaries rather than those from the opposite religion.

Mimicking Another Paramilitary Group to Discredit them or Avoid Blame

Throughout the Troubles, all the armed groups disguised their involvement in violent actions. Prior to the ceasefires in the 1990s this was because: the target was a coreligionist civilian; an operation had gone wrong resulting in a higher number of casualties than anticipated; they targeted the wrong victim; the action was not sanctioned by the leadership; there was collusion with the security forces; and the action was more of a criminal than a political act. During the ceasefires in the 1990s, it was to avoid losing political concessions gained from the peace process or even being expelled from the political talks.

False Flags

The most widespread evidence of aggressive mimicry by groups was in the use of false flags. This is when an armed group establishes a separate identity by adopting a pseudonym to carry out particular actions or forms a new group whose membership is made up of members of the parent group. This tactic is more particular to the Loyalists than Republicans. Between 1973 and 2006 the Ulster Freedom Fighters (UFF) was a cover name or false flag used by the UDA when they killed civilian Catholics. As one senior commander in the UDA reported: 'It wasn't really a separate organization, as such. No, it was basically a flag of convenience for UDA men who were carrying out military operations.'[13] In a similar vein, the Red Hand Commando (RHC) was formed in 1972 and is thought to have acted as a false flag for the UVF. The incentives to operate under a false flag increased with the introduction of the Northern Ireland (Sentences) Act in July 1998 that followed the Good Friday Agreement in April 1998. This legislation allowed for the early release of paramilitary prisoners and was a key concession made by the British government to those paramilitary groups participating in the peace process. Crucially, only those organizations that were on ceasefire could benefit from the scheme. On 11 September 1998 the first seven paramilitary prisoners were released. Also in September 1998 a group called the Red Hand Defenders (RHD) appeared and claimed responsibility for a blast bomb attack that killed a police officer. In total, the RHD killed eight people (six Catholics and two Protestants) and claimed responsibility for numerous explosions and petrol bomb attacks on Catholic families. There is a widespread assumption that members of the UDA/UFF used the RHD as a false flag to avoid being blamed for attacks and jeopardizing the early release of their prisoners.

The IRA has also used false flags, but to a lesser extent. A vigilante organization called Direct Action Against Drugs (DAAD), emerged in Belfast during the IRA ceasefire in 1994. DAAD claimed responsibility for killing eleven Catholics who were all alleged drug dealers. Although the Republican Movement has always denied any connection between the two organizations, the evidence suggests that it operated as a false flag for the IRA. The IRA dominates Republican areas of Belfast. It is unlikely that a violent vigilante group could function in their territory without their knowledge and sanction, if not their direct cooperation. Apart from DAAD, most of the vigilante groups in Republican areas were disorganized and transient. DAAD'S relative longevity of four years, suggests the IRA's ease with its shows of strength and that the two organizations were, at the very least, closely linked.

Code Words as an Identity Signal

In 1966, the UVF issued a statement to the *Belfast Telegraph* newspaper claiming that three weeks earlier it had murdered John Patrick Scullion, a Catholic.

---

[13] Interview with member of the UDA, Belfast, 15 February 2008.

Until this declaration, the police believed that Scullion had fallen over when drunk and had died from his injuries. The police exhumed Scullion's body and discovered an abdominal wound caused by a gunshot. The intriguing thing about the UVF statement was not that it revealed police ineptitude but that it included the words 'Captain William Johnson'—a code word for the UVF. From that point onwards, newspapers reported the use of code words by paramilitary groups in warnings of impending attacks, claims and denials of responsibility for acts of violence, or publicizing a policy or strategic decisions taken within the organization. These communications were mostly via phone calls made directly to the police or a third party. Media outlets, charities like the Samaritans, hospitals, and other organizations that have staff on duty twenty-four hours a day, and will have someone to receive the phone call, have all been used.

These phone calls are particularly significant if they warn of an impending bomb attack. The IRA, whose chief terror tactic was its bombing campaign, provided warnings only in the case of an economic target where civilians might be at risk. They did not forewarn if the target was a political or military one. The IRA has repeatedly pointed to their warnings as proof of their genuine desire to prevent the loss of civilian lives. However, if the receiver does not know the caller personally then they have no way to judge whether the caller is genuine. Most of this activity occurred between the late 1960s and late 1990s, in a pre-mobile phone era. The majority of calls came from a public phone box, so it was impossible to trace the identity of the caller.

There have been numerous false or hoax calls. In 1992, for example, the British transport police received 1600 hoax calls in a twelve-month period warning that bombs had been planted. Furthermore, the number of hoax calls increased after an explosion. In February 1991, British Rail and the London Underground received eighty-five bomb threats in the week following an IRA bomb at Victoria Station that killed one person. To emphasize the difficulty in distinguishing whether the threat is genuine or a hoax, the Victoria Station bomb warning was received amongst a score of hoax calls made after an explosion at Paddington Station three hours earlier (Israelson 1991).

Hoax calls fall into two broad categories. The first is when *both the caller and the message are false.* This can take one of two forms. First is when the caller has no paramilitary connections and is motivated by malice or mischief-making. The second is when the caller is a representative of a paramilitary group but is mimicking another paramilitary group to avoid taking responsibility for their actions or to discredit the mimicked group. For example, the dissident Republican groups (the CIRA and the rIRA) have mimicked the IRA in phone calls claiming that the IRA was responsible for attacks that were carried out by the CIRA and the rIRA. These claims were intended to discredit the IRA and disrupt the peace process. The second type of hoax call is when *the caller is genuine, but the message is false.* This is when a paramilitary group falsely warns of an impending attack or makes false claims regarding an event. A hoax bomb threat, which closes a town centre or cripples a transport system, can cause severe economic damage. For example, the estimated total cost of three IRA bomb scares in England in April 1997, including the cancellation of the Grand National horse race, was 250 million pounds (Ellin 1997). The IRA repeatedly used this strategy of causing maximum disruption with no loss of life. Hoax bombs also allow paramilitaries to monitor the security response to an imminent attack. They reveal details of what routes the police take to attend the incident; their response time; how many and what type of personnel are deployed; what vehicles they use; what size area around the incident site they cordon off; whether and when the Army accompanies them, and so forth. This information is used to plan future operations. Hoax bombs are also a way to distract the security forces and mask an even bigger action. McEvoy (2001, 56–57) records an aborted IRA attempt to escape from Crumlin Road Prison by driving a digger with a bomb in the front bucket through the exterior wall of the prison. The IRA had devised an elaborate series of hoax and real bomb threats at strategic locations throughout Belfast to stretch and distract the security forces to facilitate the escape.

Code words are used in the context of contested and competitive claims where paramilitary groups mimic one another and misinformation abounds. The following sections of this chapter will explore the evolution of code words as an identity signal and assess their usefulness in enabling the police and security forces to screen genuine calls from mimics and pranksters.

Data on Code Words

Evidence on code words has been difficult to gather, and this analysis draws on newspaper and interview data. The primary data source is the Nexis UK newspaper archive which was searched using the terms 'code word', 'IRA', 'UDA', and 'UVF'. This search yielded 332 separate incidents in which code words were used between 1973 and 2011. Each of these incidents was coded for the name of the paramilitary organization deemed to be responsible, message content, event type, receiver of the message, contested claims, and code word used. During this content analysis, I read an article in the *Birmingham Mail* newspaper referencing a log book in which the newspaper's telephonists recorded every threatening phone call they received between November

1971 and June 1992. The *Birmingham Mail* kindly gave me permission to analyse this log book. There were 401 calls recorded, and each call was coded for accent of the caller, where the call came from (call box, private phone,) gender of the caller, message content, whether a code word was used, and whether it was linked to a particular incident. I also interviewed two former members of the IRA and two former members of the UDA, four PSNI officers working in Northern Ireland, and two Metropolitan police officers, a former British Military Intelligence officer, a former Special Branch officer, one journalist who worked for the *Belfast Telegraph*, and an email exchange with an investigative journalist who has worked extensively in Northern Ireland.

Creation and Evolution of Code Words as an Identity Signal

There is evidence, predating the Troubles, of paramilitaries signing their actions. In the 1920s, individual IRA commanders signed their statements but stopped doing this when the conflict escalated in the late 1960s and the penalty for IRA membership increased. Even well-known Republicans, like Sinn Féin President Gerry Adams, used pen names. In the mid-1970s, while interned in prison, he smuggled out a series of articles that were published in *Republican News/An Phoblacht* depicting life inside under the name of 'Brownie' (Adams 1996, 245–249). The IRA used a *nom de guerre* from the 1940s. From the late 1960s, this was 'P. O'Neill' but it was not a code word as such and even Danny Morrison, Sinn Féin's former publicity director, did not know the exact provenance of 'P. O'Neill'.

The first recorded use of a code word—'Captain William Johnson' by the UVF in 1966—predated the first act of group mimicry: when the Ulster Protestant Volunteers mimicked the operational behaviour of the IRA in 1969 to discredit them. The temporal order of an act of group mimicry and the adoption of code words is inverted. So, while the UVF may have originally decided to use a code word as an identity signature, it was not initially used to combat mimicry. The other paramilitary organizations soon followed the UVF in using code words. The UFF have used 'Captain Black' since the beginning of the Troubles (Taylor 2000, 118). The UDA used 'The Ulster Troubles', 'Crucible', and 'Titanic' as code words (Rowan 2010). The IRA used 'Double X' for the 1974 Birmingham pub bombings and they reportedly have also used 'Michael Collins' and 'Easter Rising', and continue to use the *noms de guerre* 'P. O'Neill' and 'Óglaigh na hÉireann'. The rIRA used the code word 'Martha Pope'.

According to data from the Nexis newspaper archive, code words were used in 332 separate incidents between 1973 and 2011 and they were used with the greatest frequency between 1992 and 2003. In these data, code words were also used more often by Republicans than Loyalists: thirty-five per cent of the incidents with code words were attributable to the IRA and fifteen per cent to the rIRA. Of the code words used by the IRA, fifty-eight per cent were in messages warning of an actual bomb or explosion and fourteen per cent were in messages that subsequently turned out to be hoax calls. Among the Loyalists, code words were mostly used by the Red Hand Defenders. Twelve per cent of all incidents with a code word were attributed to this UDA-linked organization. From the available data, it is clear that the IRA used code words more than any other organization, and they used code words in messages referring to both genuine and hoax incidents. The greater proportion of code words were used in calls forewarning of a bomb attack (either real or a hoax), and the IRA were the primary bombers throughout the Troubles.

Protocol for Verifying Code Words

To ascertain whether a code word is an identity signature, there needs to be a way of verifying that the code word is linked to the paramilitary group using it. There is no evidence of joint decisions or agreements between the paramilitaries and state security or media agencies about what the code words will be. Joe Mulholland, former news director at RTE commented 'We can't be meeting with these people to work out a code word, sitting down in a pub and say, "Right, let's agree on a new code word"' (Glauber 1996). In interviews, the police denied that any code words were agreed and stated 'there is no master list of code words locked away in a filing cabinet.'[14]

Code words are verified in two ways. The first is if they are linked to an event. If the IRA forewarn of an explosion using a code word, the explosion occurs, and the IRA's authorship is confirmed using additional investigative techniques, then the code word is verified as a genuine signature. For example, the head of Scotland Yard's anti-terrorist branch, Commander George Churchill-Coleman, stated that one code word was used on sixty-one occasions, each time claiming that explosive devices had been placed. On twenty-two occasions bombs exploded, two were made safe, and the remaining thirty-seven were hoaxes of a type where the caller was genuine but the message was false (Burden 1992). This method depends on security agents

---

[14] Interview with member of the PSNI, Belfast, 23 August 2012.

receiving the same code words over time and sharing information within and between security agencies.

The second way in which code words are verified is when they are communicated by a known member of the relevant paramilitary group. The IRA ended a seventeen-month ceasefire in February 1996 with a massive bomb in Canary Wharf London. This attack was very significant politically and militarily. The IRA wanted the British government to be sure that it was they who had planted the bomb. Their intent was to cause maximum economic damage but to minimize the number of civilian casualties. They therefore needed their warning to be taken seriously and acted upon. The warning call was made ninety minutes before the bomb exploded; the information about the location of the bomb was accurate, and a code word was given. However, the IRA man who telephoned the warning to a Dublin newsroom called a specific journalist whom he had spoken to before. The journalist recognized his voice, knew that he was representing the IRA, and that the warning and the code word were genuine.

Verification through a known contact did not and cannot take place in all cases. The paramilitaries did not consistently use the same individuals to make phone calls using code words. A former IRA interviewee reported that the brigadier adjutant would give out the code words and would task a lower ranked individual to make the call.[15] Lower ranked members are likely to have served in an organization for a shorter period and are thus less likely to have formed long-term relationships with intermediaries who could verify their identity. Furthermore, the paramilitaries did not systematically call the same organization or individuals; they used a plurality of outlets. The data collected from the Nexis database reveals that fourteen different types of outlet including various types of media (radio, television, and newspapers), charities, and named individuals received calls. Newspapers received twenty-nine per cent of the calls followed by television stations (fourteen per cent), radio stations (ten per cent), and the Samaritans (three per cent). Individual journalists, such as Henry McDonald from the *Guardian* and Brian Rowan from the *Belfast Telegraph*, claim to have a special relationship with paramilitaries that is facilitated by code words (see McDonald 2011; Rowan 2011), but most journalists staffing a newsroom did not.[16] The inquest held in the aftermath of the 1998 Omagh car bomb planted by the rIRA which killed twenty-nine people (and two unborn twin girls) and injured hundreds of others revealed that staff at one of Northern Ireland's main television stations (UTV) who

---

[15] Interview with former IRA member, Belfast, 23 August 2012.
[16] Telephone interview with *Belfast Telegraph* journalist, Oxford, 10 September 2012.

received a telephone warning thirty minutes before the explosion did not have a list of recognized code words, despite regularly receiving such warnings. The phone operators at the *Birmingham Mail* also had no knowledge of any code words.

Are Communications Containing a Code Word Taken More Seriously?

In 1991–1992 British Transport Police received about 4000 bomb warnings believed to be part of an IRA campaign to disrupt British Rail and the London Underground (Clutterbuck 1997). The police did not have the resources to deal with such a large number of bomb threats. They had to judge which calls to investigate and which to ignore. Of those 4000 warnings received by the British Transport Police, in ninety-eight per cent of cases, the police let traffic continue uninterrupted. They closed a station or a line on only seventy-two occasions (under two per cent), and there was a bomb on twenty-five of these occasions (fourteen explosive and eleven incendiary) (Clutterbuck 1997). If code words were identity signatures, we would expect the police to trust messages containing a code word more than those without a code word, and we might also expect their response to be different. The security personnel who were interviewed said they assess the presence or absence of a code word alongside other notable information such as the accent of the caller, their gender, (pre-mobile phones) whether the call came from a call box, the content of the rest of the message, and whether the warning conformed to any pattern of previous warnings received. They also consider any intelligence about planned paramilitary activity. The interviewees unanimously stated that the police take all calls warning of imminent attacks equally seriously. They argued that their response was the same regardless of the presence of a code word. They cannot risk the catastrophic consequences of mistakenly assuming that a warning without a code word is a fake. However, despite this they also acknowledged that a code word increased the 'sense of urgency' when assessing how real the threat of attack.

This evidence seems contradictory, but it is nonetheless clear that the presence of a code word alone does not determine the police response. In 1997, organizers postponed the Grand National horse race and evacuated 70,000 people from the racecourse after police received a coded bomb warning which turned out to be a hoax. Coded bomb warnings were again received from the IRA before the rerun of the race two days later. However, the police and organizers decided that the race should go ahead, and it was attended by the then-prime minister, John Major. Once more the calls were hoax calls.

Code words were used not just to warn of bombs but also in claims and denials of responsibility for acts of violence. Here, too, the police use 'good old fashioned police detective work' to look for additional evidence of who authored the act, such as the timer on the explosive device.[17] According to one interviewee, only the IRA used Swiss watch timers from Libya.[18]

So we can conclude that code words are identity signatures but they are not robust. They are evidence of authorship only in conjunction with other identifying information. The difficulty in verifying their authenticity means that they can easily contribute to an act of mimicry. In a phone call, mimics can claim to be a paramilitary and issue a bomb warning with a 'code word'. A police investigation into London Underground found employees inciting others to make bomb threats using code words received in threatening calls to their office (Hunter and Bhatia 1994). When the police receive a bomb warning, they must assess whether the identity of the caller is genuine and whether the warning is real or a hoax. Paramilitaries have used code words to persuade the police that hoax calls are real bomb warnings to cause, among other things, economic disruption. The use of code words in hoax calls makes code words less trustworthy and undermines their ability to communicate reliable information about the veracity of the warning.

Conclusion

This research empirically explores the role of defensive and aggressive mimicry by individuals and groups in the Northern Ireland conflict. Protestants and Catholics share physical features and have many opportunities to observe one another, yet there are few cases of defensive mimicry, where a member of one religion mimics a member of another religion to avoid harm. The strong group boundaries that facilitate high social knowledge of who belongs to what religion constrain its use (see also Laitin, this volume). Aggressive mimicry, where the model is an individual, is also less prevalent than we might expect. We do observe mimicry when the assassins need to manoeuvre themselves or the victim into a better position to kill them; when assassins need to persuade their victim to accompany them to another location; when assassins need an extra element of surprise to kill their victim; and when there is an additional pay-off to the act of mimicry. The analysis has also shown how a single act of mimicry by individuals and groups can be deployed for both aggressive and defensive purposes.

---

[17] Telephone interview with member of the PSNI, Oxford, 20 April 2012.
[18] Interview with former Special Branch officer, London, 19 April 2012.

The two primary tactics employed by the paramilitary groups—bombings and targeted assassinations—could be accomplished without mimicry. Bombers who shared physical and behavioural traits with the opposite religion could plant explosive devices while remaining camouflaged. They did not need mimicry to go undetected. Murders of members of the opposite religion often required the assassins to go into enemy territory. They needed to kill and retreat as quickly as possible. Stephen Holmes (this volume) describes how mimicry allows an assassin to get a 'momentary jump' on his victim. However, the shared physical characteristics between Protestants and Catholics meant that victims could easily mistake attackers for a non-threatening third party or a member of the victim's group. Mimicry was unnecessary. Also, camouflage or blending in rather than mimicry was their best escape strategy. In general, not wearing any identifiable clothing like a uniform or any item that singled them out aided their escape (the exception to this being prison escapes).

Aggressive mimicry appears more prevalent at the group level than at the individual level. There were instances of group mimicry throughout the Troubles and the incentives to mimic other groups increased dramatically following the Good Friday Agreement in 1998. Paramilitary groups mimicked each other to carry out assassinations, discredit other groups. and to avoid blame for acts of violence. They also operated under false flags.

A reduction in the amount of paramilitary violence has also led to the emergence of a different category of aggressive mimicry perpetrated by criminal extortionists who mimic being paramilitaries. As paramilitary control diminishes, and they are less able to punish those free-riding on their reputation for violence, so that free-riding has increased. Victims are particularly vulnerable if they have poor-quality information about Northern Irish paramilitaries and are duped by criminals displaying the signs or 'trademarks' of being a paramilitary (see also Mamidi, this volume). In May 2007, a criminal gang extorted £820,000 from a partner in DHK & Co Solicitors firm in London. The gang claimed to represent the IRA. To their victim, who was of Indian origin, they carried all the signs of IRA—they had Irish accents, and they issued death threats against him and his family (Lattice 2008). Perhaps a savvier victim would have distinguished the gang leader's Southern Irish accent as being different from the predominantly Northern Irish accents of IRA members and questioned their claims. But persuaded that they were the IRA and terrified by the IRA's reputation for violence, he appropriated the money and caused the collapse of his law firm. Few of these cases come to the surface. If they are successful, we may not know about them but the incentives to free-ride on a paramilitary group's reputation for violence in a context of low-quality information exist.

This chapter also explores the evolution of code words as an identity signature. They are used by paramilitaries when they forewarn about explosions, claim, or deny acts of violence, and communicate their policy or strategic decisions to a wider audience. Code words evolved into an anti-mimicry device in the context of paramilitary groups mimicking one another and contested and competitive claims as to the authorship of acts of violence. They also evolved in a wider context of an abundance of poor-quality information circulating in the public and security domain about the authorship of acts of violence, and in an environment where information was not shared easily between various parts of the security services.

The IRA, in particular, used code words in their bombing campaign to forewarn of real bombs and in hoax bomb warnings intended to cause maximum economic disruption. It is important to be clear that the code words are designed to authenticate the identity of the caller and not the veracity of the message. However, the false content of the messages has undermined the credibility of the code words. The fact that the paramilitaries sent messages via a plurality of third parties also restricted the ability of code words to be a credible identity signal. Often code words were misheard or misunderstood and the content of the message was passed on incorrectly.

In conclusion, the evidence points to paramilitaries and police viewing code words differently. There is evidence that the paramilitaries considered code words to be a stand-alone identity signal. But, for the police, code words were only one part of a more complex identity signal that was comprised of information about the accent and gender of the caller, the location of the call, the full content of the message, whether the warning conformed to any pattern of previous warnings received, and intelligence about planned paramilitary activity.

References

Adair, Johnny. 2007. *Mad Dog*. London: John Blake.

Adams, Gerry. 1996. *Before the Dawn: An Autobiography*. Dingle: Brandon.

AnPhoblacht. 1997. 'How the Audacious Escape Was Done. First Interview with Liam Averill'. *AnPhoblacht/Republican News*, 8 January.

Belfast Telegraph. 2001. 'RUC Probe '92 Murder of Scottish Catholic'. *BelfastTelegraph.co.uk*, 14 January.

Belfast Telegraph. 2007. 'Nairac: An Undercover Hero or a Maverick Fool?'. *Belfast-Telegraph.co.uk*, 13 May.

Benmelech, Efraim and Claude Berrebi. 2007. 'Human Capital and the Productivity of Suicide Bombers'. *The Journal of Economic Perspectives (Nashville)* 21 (3): 223–238.

Burden, Peter. 1992. 'Thursday Six Hurt in IRA Blitz; Anti-Terrorist Chief Hits out at Deadly Games after Bombs Blast Railways and TA Centre'. *Daily Mail*, 22 October.

Campbell, Colm and Ita Connolly. 2006. 'Making War on Terror? Global Lessons from Northern Ireland'. *The Modern Law Review* 69 (6): 935–957.

Clutterbuck, Richard. 1997. 'Ban the Bomb Chaos; Should We Be Held up for Hours by Police Acting on a Coded Telephone Warning? Richard Clutterbuck Thinks It May Be Time for Us to Be Allowed to Take Our Chances'. *Guardian*, 23 April.

Coogan, Tim Pat. 1995. *The Troubles: Ireland's Ordeal 1966–1996 and the Search for Peace*. London: Hutchinson.

Coogan, Tim Pat. 2000. *The IRA*. London: HarperCollins.

Ellin, Tony. 1997. 'Travel Chaos as IRA Gang Strike Again; The IRA Tried to Bring Britain to a Standstill Yesterday with Two Small Bombs and a String of Hoax Calls'. *Daily Record*, 19 April.

Gambetta, Diego. 2005. 'Deceptive Mimicry in Humans'. In *Perspectives on Imitation: From Neuroscience to Social Science*. Volume 2: *Imitation, Human Development and Culture*, edited by Susan L. Hurley and Nick Chater, 221–242. Cambridge, MA: MIT Press.

Gambetta, Diego and Heather Hamill. 2005. *Streetwise: How Taxi Drivers Establish Their Customers' Trustworthiness*. New York: Russell Sage.

Glauber, Bill. 1996. 'Frightful Words No One Wants to Hear: IRA Speaks in Code When It Warns Press of Bombs, Attacks'. *Baltimore Sun*, 11 October.

Hamill, Heather. 2011. *The Hoods: Crime and Punishment in Belfast*. Princeton, NJ: Princeton University Press.

Hatterstone, Simon. 2001. 'Bombs and Books'. *Guardian*, 10 December.

Hearst, David. 1988. 'Fake Police Kill IRA Man'. *Guardian*, 26 July.

Hewstone, Miles, Ed Cairns, A. Voci, S. Paolini, A. McLernon, R. Crisp, U. Niens, and J. Craig. 2004. 'Intergroup Contact in a Divided Society: Challenging Segregation in Northern Ireland'. In *Social Psychology of Inclusion and Exclusion*, edited by Dominic Abrams, Michael A. Hogg, and José M. Marques, 265–292 New York: Psychology Press.

House of Commons. 1998. 'Report of an Inquiry into the Escape of a Prisoner from HMP Maze on 10 December 1997 and the Shooting of a Prisoner on 27 December 1997'. London: HMSO.

Hughes, Joanne, Andrea Campbell, Miles Hewstone, and Ed Cairns. 2007. 'Segregation in Northern Ireland'. *Policy Studies* 28 (1): 33–53.

Hunter, Flora and Shekhar Bhatia. 1994. 'Train Bomb Hoaxes: LT Official Held'. *Evening Standard*, 7 January.

Israelson, David. 1991. 'U.K. Bomb Threats Cause Transit Chaos'. *Toronto Star*, 22 February.

Kalyvas, Stathis N. 2006. *The Logic of Violence in Civil War*. Cambridge: Cambridge University Press.

Kelly, Gerry. 2013. *The Escape. The Inside Story of the 1983 Escape from Long Kesh Prison*. Belfast: M&G Publications.

Lattice, Tony. 2008. 'London Solicitors Firm Fall Victim to Alleged Bogus IRA Extortion Racket', 25 January. http://www.lawyersolicitor.co.uk/news-press-releases/ext/london-solicitors-firm-fall-victim.1 (accessed 23 March 2015)

Lister, David and Hugh Jordan. 2003. *Mad Dog: The Rise and Fall of Johnny Adair and 'C Company'*. Edinburgh: Mainstream.

McDonald, Henry. 2011. 'Police Must Accept Journalists' Confidential Sources Are Sacrosanct'. *Guardian*, 23 September.

McEvoy, Kieran. 2001. *Paramilitary Imprisonment in Northern Ireland: Resistance, Management, and Release*. Oxford: Oxford University Press.

McKittrick, David. 1997. 'Murder Verdicts Quashed in Ulster Sectarian Killing'. *Independent*, 22 March.

McKittrick, David, Seamus Kelters, Brian Feeney, Chris Thornton, and David McVea. 2007. *Lost Lives: The Stories of the Men, Women and Children Who Died as a Result of the Northern Ireland Troubles*. 2nd edn. Edinburgh: Mainstream.

Moloney, Ed. 2010. *Voices from the Grave: Two Men's War in Ireland*. London: Faber & Faber.

Rowan, Brian. 2010. 'Growing up with the UDA'. *Belfast Telegraph*, 11 January.

Rowan, Brian. 2011. 'Reporters Cannot Become State Evidence-Gatherers'. *Belfast Telegraph*, 27 August.

Rutherford, Adrian and Carla Bowyer. 2011. 'Infamous Peace Line That Cut Public Park in Two Opens after 17 Years'. *Belfast Telegraph*, 17 September.

Sluka, Jeffrey A. (ed.). 2000. *Death Squad: The Anthropology of State Terror*. Philadelphia, PA: University of Pennsylvania Press.

Spjut, R. J. 1986. 'Internment and Detention Without Trial in Northern Ireland 1971–1975: Ministerial Policy and Practice'. *The Modern Law Review* 49 (6): 712–739.

Taylor, Peter. 2000. *Loyalists*. London: Bloomsbury.

Urban, Mark. 1992. *Big Boys' Rules: The SAS and the Secret Struggle against the IRA*. London: Faber & Faber.

Weitzer, Ronald John. 1995. *Policing Under Fire: Ethnic Conflict and Police-Community Relations in Northern Ireland*. Albany, NY: State University of New York Press.

5

The Red Brigade Signature

Mimic-proofing Claims of Political Violent Actions, Italy, 1969–1980

Valeria Pizzini-Gambetta and Diego Gambetta

In this chapter we investigate how groups of both left- and right-wing violent extremists claimed their attacks between 1969 and 1980 in Italy. We discuss the various threats posed by mimics to an underground group's communication strategies, describe the measures groups adopted to protect themselves against these threats, and how their mimic-proofing strategies shaped their development. We will focus in particular on the most prominent left-wing violent organization known as *Brigate Rosse* (BR), paying attention to the techniques they employed for signing their violent acts and asking two questions: whether and how BR's 'signature' was designed in such a way as to prevent mimicry, and whether they were successful at that.

Political violence by clandestine groups has a strong communicative component. The success in promoting their cause depends upon the publicity that their deeds receive (Lacqueur 1997, 110). The violent action is instrumental to promoting a cause that they cannot fight as a conventional war 'in the light of day [where] there is no mystery about the identity of the participants' (Lacqueur 1997, 3). Perpetrators of political violence want for obvious reasons to keep their personal identity and whereabouts secret, but at the same time they want their group to take the credit for what they do. This explains why they often go to great lengths to claim responsibility for their actions, and may develop a 'symbiotic' relationship with the media (Enders and Sandler 2006, 37; Gambetta 2005a; Hoffman, Shelton, and Cleven 2013; Rohner and Frey 2007; Wilkinson 1997).

The notion of political violence as 'propaganda by the deeds' predicts that, given their effort and investment of resources into an act of violence, perpetrators will claim paternity of that action as 'a private good' not to be shared with other groups (Enders and Sandler 2006, 37; Kydd and Walter 2006).

Valeria Pizzini-Gambetta and Diego Gambetta, *The Red Brigade Signature*. In: *Fight, Flight, Mimic*. Edited by: Diego Gambetta and Thomas Hegghammer, Oxford University Press. © Valeria Pizzini-Gambetta and Diego Gambetta (2024).
DOI: 10.1093/9780191802454.003.0005

How clandestine groups manage to link their violent political actions and the identity of their group has been analysed as a simple choice between to claim or not to claim responsibility (Hoffman 2010; Min 2013; Rapoport 1997; Wright 2011). The studies that have investigated this choice have identified five conditions, each a sufficient reason that makes claiming the preferred option; when groups want to:

(i)   reach a wide audience and inspire them to support their cause (Sánchez-Cuenca 2007)
(ii) compete with other groups to represent a cause (Hoffman 2010)
(iii) signal their strength (Wright 2011)
(iv) sustain a strategy of intimidation (Enders and Sandler 2006; Min 2013)
(v)   provoke a government to launch a repressive reaction with the aim of radicalizing constituencies at the expenses of moderate factions (Bueno de Mesquita and Dickson 2007).

Only if they pursue none of these aims do they prefer to shroud their actions in anonymity.

These studies, however, consider as unproblematic the manner in which the perpetrators of violence communicate their group identity and link it to their actions. Yet, matters are not so simple. Well before the Internet made trusting the identity of one's interlocutor even more problematic (see Hegghammer, this volume), signalling the identity of clandestine political groups faced issues of credibility: how does one know beyond reasonable doubt who is the author of a claim of responsibility? How does one trust that a claim is truly made by the group whose name is being spent, or whether the group is the real perpetrator of the claimed attack?

One cannot just throw a bomb or kneecap someone and effortlessly announce it to the world. 'Property rights' over actions are often contested. The true perpetrators have to contend with false claims issued by competitors in their own political camp who want to free-ride on the reputation of their actions, or with false flag operators of the opposite camp who commit acts of a kind aimed to discredit them and forge their signatures. To protect themselves from these perils and avoid harmful intrusions into their communicative domain, groups need to take a variety of actions, above all design a unique composite of identifying markers which will form their 'signature'. A name or an acronym without supporting evidence does not suffice for an underground political group to establish the authenticity of their claims and becomes fair game for mimics. Everybody can make a phone call, and voice a name. The groups that want to maintain exclusive 'property rights' over their actions must ensure that receivers perceive their claims as credible. Signalling theory stipulates that credibility of a signal is established if the receiver believes that only the true perpetrators can afford the cost of emitting that particular identity signal (Bacharach and Gambetta 2001; Gambetta 2005b, 2009a, 2009b; Spence 1973).

The Italian landscape of political violence offers ample variations in the range of claims attached to acts of political violence: it goes from no claim at all—chosen by groups at both ends of the political spectrum that kept a legitimate façade while running an underground armed branch (Lotta Continua, Potere Operaio, Autonomia Operaia, Movimento Sociale Italiano)—to umbrella names identifying clusters of political agents loosely united by common aims rather than as members of an actual organization (Nuclei Armati Rivoluzionari, Anarchists), to names identifying a properly structured group—such as the BR and others (Nuclei Armati Proletari and Prima Linea). This chapter deals only with the third type of claims, and with the three types of mimics that they attracted: *criminals*, who exploited the names of violent political groups to strengthen their ability to intimidate their victims or to confuse law and order agents (see Mamidi, this volume); far-right *agent provocateurs* carrying out false flag operations in order to blame them on their 'model' groups; and *left-wing militants* competing with the BR who signed with their names actions they did not carry out. Before proceeding, however, we need to say a few words about the origins of the political conflict in which the BR were involved.

How It All Began

Between 1969 and 1980 Italy witnessed a wave of political violence, which caused the highest number of victims by an ideologically driven conflict in Europe after the Second World War: 362 dead (150 caused by bombings) and 673 wounded (551 caused by bombings) (Della Porta and Rossi 1984; Galleni 1981; Moss 1989, 18–19). The violence reached its apex in 1978, both in terms of number of attacks[1] and type of targets: on 16 March 1978, in Rome, the BR kidnapped Aldo Moro, the president of the Christian Democratic Party, and fifty-five days later assassinated him, after the negotiations with the Italian government failed.

---

[1] According to the Italian Ministry of Interiors, in 1978 there were 2498 acts of political violence (Moss 1989, 19).

The violence was not due to a military confrontation, it was rather the peak of a thirty-year long simmering conflict in which 'terrorism has been more or less a permanent phenomenon' (Engene 2004, 134). The Resistance against the Fascists and the Germans in Italy (1943–1945) was not only a war of liberation but a civil war (Pavone 1991), and as such left a trail of bitter grievances which continued to feed political tensions and often violence. The winning side, made of former fighters loyal either to the Italian Communist Party or to a centrist and liberal coalition, harboured contrasting international loyalties while sharing a great deal of resentment towards the soft treatment of the losing side. The losers, unified in defeat, were minimally penalized by the newly formed Italian Republic—a large section of the Fascist secret police, for instance, was entrusted with state security (De Lutiis 1996; Pisano 1979).

The anti-communists and the far right, both within and outside state institutions, played a role in the Italian branch of the 'Stay Behind' network, which was organized by the Allied security services to oppose the expansion of Communist influence over Western Europe (Ganser 2005; De Lutiis 1996). The fear of the 'red danger' was fuelled not just by propaganda, but by covert acts of political violence; it is, to this day, unclear to what extent these operations were conceived with or without the approval of the 'Stay Behind' chiefs (Bale 1996; Ferraresi 1996). 'Intoxication techniques', effectively false flag operations, became part of this strategy in the 1960s and aimed at fomenting popular support for a political turn to the right should the risk of a government including the Communist Party in Italy ever arise (Bale 1996; Ferraresi 1996; Panvini 2009; Weinberg 1995). According to Stefano Delle Chiaie, the founder of the radical right-wing formation, Avanguardia Nazionale, and one of the main promoters of this campaign, 'the main element of the strategy of destabilization was the creation of fake radical left-wing groups and the infiltration of pre-existing groups so that the left would be held responsible of the acts of political violence [...] thus provoking the army to intervene and exclude the Communist Party from any influence in Italian politics'.[2]

In 1968, after social tension had mounted for a few years and center-left governments were being experimented, that risk did indeed increase: the response was a bombing campaign, which culminated with the bombing of the Banca dell'Agricoltura in Milan in December 1969, which killed seventeen people and wounded eighty-eight (Bale 1987, 211, 1996, 156; De

---

[2] Tribunale di Milano, 1995. Ufficio Istruzione GI Giovanni Salvini, Sentenza Ordinanza vs Azzi Nico et alii, n.2643/84 RGPM, Testimonianza di Stefano delle Chiaie, 16.6.192, fogli.3–4.

Palo and Giannulli 1989, 127).[3] This outrage, and similar ones that followed, as well as the murky and inconclusive investigations of their perpetrators (Barbieri and Cucchiarelli 2003), became a focal point for the years of political violence that erupted in the 1970s.

The bombing of the Banca dell'Agricoltura—the critical event that set in motion the wave of political violence that followed—was itself an instance of deceptive mimicry. The perpetrators, while not signing their attacks, manufactured clues that pointed to left-wing extremists: within a few hours the police hastily attributed the attack to an anarchist cell, the Circolo XXII Marzo (Ferraresi 1996, 92; De Palo and Giannulli 1989, 152). This group was a recent spin-off of a long-established Anarchist group—a motley crew reminiscent of the fictional anarchists in Joseph Conrad's *The Secret Agent*, which included 'comrade Andrea', a police undercover agent, and Mario Merlino, a member of the far-right.[4] As Merlino himself later confessed, he was an infiltrator 'specialized in left-wing groups' (De Palo and Giannulli 1989, 145) and part of a network of right-wing extremists linked with US military agents in northeastern Italy. After unsuccessfully trying to infiltrate Maoist groups, he instigated the formation of Circolo XXII Marzo (Ferraresi 1996, 92). To look credible he grew long hair, donned 'leftist' appurtenances, and displayed anarchist symbols.

Merlino's case is an instance of an individual mimic infiltrating a political group, and his experience shows why the Anarchist movement, unconcerned with its identity marks, was such a soft target for infiltration (De Palo and Giannulli 1989, 147–148): their simple logo was easy to reproduce; unlike what happens in centralized groups, there was no or little monitoring of the credentials of those who claimed to belong to it; its loose ideological doctrine made access easier for impostors with a scant knowledge of Marxism who would have failed to convince members of stricter groups. Furthermore, anarchists' typical targets were mostly banks and churches, and Fascist infiltrators, who wanted to prove to be bona fide anarchists, could participate in the attacks without violating their political loyalty. Left-wing groups instead were not such easy models to mimic, they hit mostly right-wing targets, thus raising the cost of accreditation for Fascist infiltrators to a level they were not willing to pay.

[3] Between 1968 and 1969 there were attacks in Rome, Reggio Calabria, Reggio Emilia, Terni, Legnano, and Milan signed with the Anarchist symbol or saluting Chairman Mao that were later identified as the work of right-wing extremists (De Palo and Giannulli 1989, 150–151); also (Ferraresi 1996).

[4] Other members were Pietro Valpreda, a middle-aged dancer, three teenage students of middle class origins, and a university student (De Palo and Giannulli 1989, 152).

The 1969 bombing in Milan brought about a dramatic radicalization. It reinforced the expectations of left-wing militants that violence was a necessary defensive option (Ferraresi 1996), it was a 'watershed' (Grandi 2005, 260), 'a crucial step in the individual paths towards terrorism' (Cazzullo 2006, 90–99; Franceschini, Buffa, and Giustolisi 1988, 23; Gallinari 2009, 73–74; Grandi 2005, 26, 61, 112, 193; Moretti 2000, 11). Francesco Pardi, a senior member of Potere Operaio in Florence, said:

> it radically changed our politics and how we perceived our adversary. Before [the Banca dell'Agricoltura bombing] our antagonists were individual social actors [...] the bombing [...] and in particular the unacceptable behavior of those who represented the state at the time, convinced us that the state was our enemy (Pardi in Grandi 2005, 282).

Individual paths to political violence converged into organized groups, first among them the BR (Moretti 2000, 20).[5] A major act of mimicry was arguably the original sin that set in motion this wave of political violence.

The chapter is organized as follows. First, we present our data and sources, and next we give a quantitative overview of claimed political actions. Third, we describe the basic features of the names used to claim, and the puzzling sizeable presence of claims of responsibility that use names which appear only once. Fourth, we give an overview of false claims and of their types, and describe the identity and the motives of the mimics who issue them. Fifth, we analyse how the groups that are victims of mimicry or simply fear it, respond whether ex-post, by issuing disclaimers and counterclaims, or preventively by designing a mimic-proof signature. In the last section, we focus on how and why the BR came to invent their name and logo, and to design their signature.

Data and Sources

In order to illustrate the reasoning that led to the decision to claim and to the various forms of signatures, we draw information from the biographies and autobiographies of former militants and from the judicial papers of the

---

[5] Similarly, Alberto Franceschini, one of BR founders, claims: 'the bombing of] Piazza Fontana confirmed the need and the virtue of our project. If we needed something that would strengthen our resolve and would tell us that there was no time to waste, well, the carnage in Piazza Fontana was exactly that' (Fasanella and Franceschini 2004, 56 our translation).

trials of extremists. This large anecdotal material was particularly useful for describing how the BR planned their communication strategies.

The main source we used, however, is the collection of violent events provided by the Global Terrorism Database (GTD) of the National Consortium for the Study of Terrorism and Responses to Terrorism (START), expanded and enriched by a systematic search of Italian primary sources. GTD is widely used by scholars and considered the most comprehensive open-source database on terrorist events worldwide (LaFree 2010). It is event-based and the version we used included in total about 98,000 acts of political violence between 1970 and 2010.[6] For Italy, GTD records fewer violent events than those recorded by the Italian Ministry of Interior and by the Italian Communist Party, which closely monitored political violence (Della Porta and Rossi 1984; Galleni 1981; Moss 1989). The trends, however, follow a similar trajectory, regardless of the differences in absolute numbers between the three sources (Figure 5.1).

The GTD provides some detailed information for each violent event (weapons, target, victims, attribution etc.). However, for the events of the period under study, it does not distinguish between third-party attribution of responsibility and perpetrators' claims; it gives scant details on the manner of the claims, and false claims are not identified. Therefore we integrated GTD by searching each daily issue of the Italian newspaper *Corriere*

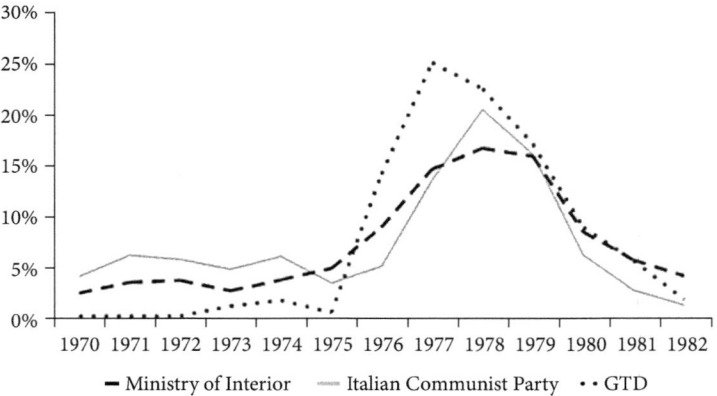

Figure 5.1 Frequency distribution of the total number of acts of political violence by year and source, Italy, 1970–1982
Source: Ministry of Interior and Italian Communist Party quoted in Moss 1989, GTD database

[6] The version we used was downloaded on 26 April 2012.

della Sera (CdS).[7] Whenever available, we added to each event in the GTD database, whether the act was the result of riots or a planned operation, whether it was claimed or only attributed to a group, whether there were contentious claims, disclaimers, or counterclaims, and if mimics were detected around the time of the event. We also recorded details of the signature techniques. We supplemented the database using dedicated websites,[8] a host of sources that collect and classify acts of political violence in Italy (Galleni 1981; Marletti et al. 2004; Progetto Memoria 2007), and the biographies of some of the protagonists. These additional sources offered data about false attributions and mimics that became apparent over time following judicial investigations. Based on all this information, we added a series of variables for the events in the GTD database, and thus created a new database, the GTD-CdS.

We chose to restrict the focus to the period 1974–1980 for three reasons. Firstly, because between 1974 and 1980 there was an anomalously high wave of political violence in Italy: seventy-three per cent of the violence episodes recorded in GTD over forty years (1108 out of 1509), occurred in just those six years. Secondly, the BR became a structured clandestine group around 1974 and split into three independent groups in 1981. Finally, between 1969 and 1974, political violence in Italy was mostly anonymous (Galleni 1981), and as such it required a detailed analysis that is beyond the scope of the present chapter.

During this search we also uncovered 715 other violent acts, which while not included in the original GTD database, matched the criteria for inclusion according to its code book.[9] These are mostly minor attacks, yet a majority of them were claimed indicating that they had communicative value for their perpetrators, and we therefore chose to add them. As a result, the combined GTD-CdS database now records 1823 events between 1974 and 1980.

---

[7] *Corriere della Sera*, based in Milan, is the oldest Italian newspaper, has the largest readership and broadest coverage of national events. *CdS* covered extensively the political violence in northern Italy, and in particular in Milan and Turin, two out of three main theatres of political violence in Italy in the period.

[8] The online sources are: GTD: http://www.start.umd.edu/gtd/ (downloaded 26 April 2012); Associazione Italiana Vittime del Terrorismo: http://www.vittimeterrorismo.it/ (last accessed 4 May 2015); Red Brigades Historical Archive: Brigaterosse.it (last accessed March 2008, the site is no longer available online); and the Archive of the Italian Anti-Fascist Movement: www.pernondimenticare.net (last accessed 4 May 2015)

[9] We include only violent events which were planned in advance, while we do not include violence which came as a consequence of riots. Events included in GTD must: a) entail violence or the threat of violence; b) be intentional; c) must be perpetrated by a subnational actor. In addition, at least two out of the following three criteria must be present: 1) The act must be aimed at attaining a political, economic, religious, or social goal; 2) There must be evidence of an intention to coerce, intimidate, or convey some other message to a larger audience (or audiences) than the immediate victims; 3) The action must be outside the context of legitimate warfare activities (GTD 2011: 5).

Claims

In the wave of political violence that hit Italy between 1974 and 1980, three quarters of the 1823 attacks recorded in GTD-CdS database were relatively minor, damaging property rather than harming persons. Still, twenty-two per cent were serious acts of violence against people: wounding (204, eleven per cent), assassinations (108, six per cent), major bombings (sixty-four, four per cent), and kidnapping (twenty-three, one per cent) (Figure 5.2).

After a decline at the beginning of the period, the share of serious attacks against persons steadily increased, in particular after 1976, reaching thirty-five per cent in 1980 (Figure 5.3).

Of the 1823 attacks, forty-five per cent were never claimed—the perpetrators of most of these attacks remain unknown though in some cases (thirteen per cent of the total claimed) the attack was later attributed to a particular group by a third party, usually the police. However, fifty-five per cent (1012) were claimed by at least one group—by a claim we mean an oral or written statement intended to reach the public, which purports to be speaking on behalf of a group identified by a name, and which attributes to that name the responsibility of a specific violent act. The propensity to issue a claim (Figure 5.4) appears to be higher for major attacks: eighty per cent of assassinations were claimed, and kidnappings for political reasons in ninety-three per cent of cases. Even in this field the harder one works the more one wants to benefit from the fruit of one's labour.

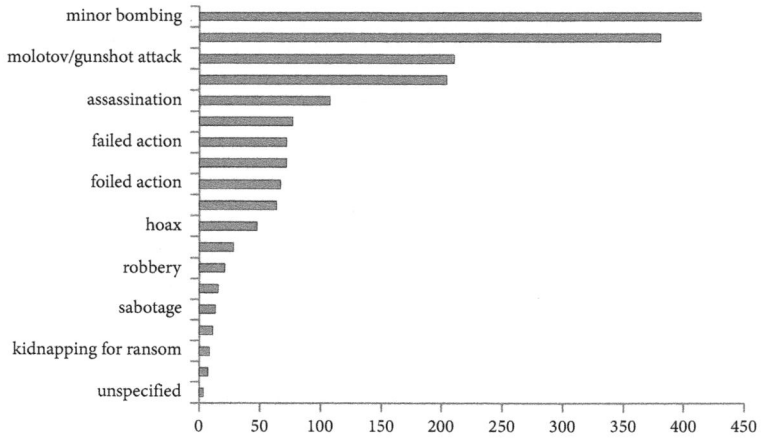

Figure 5.2 Acts of political violence by type, Italy, 1974–1980 (N = 1823)
Source: GTD-CdS database

Figure 5.3 Relative frequency by year of major acts over total acts of
political violence, Italy, 1974–1980 (N = 1823)
Source: GTD-CdS database

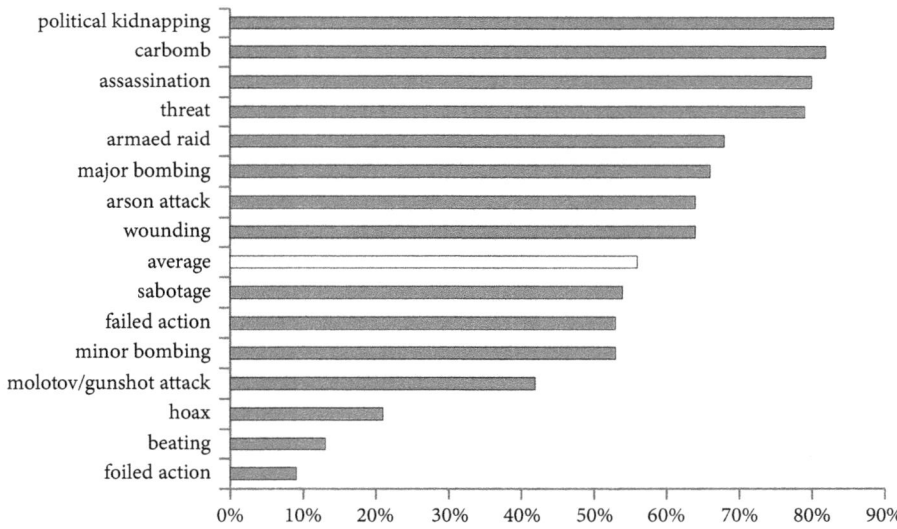

Figure 5.4 Relative frequency of claimed acts of political violence by types of attack,
Italy 1974–1980 (N = 1012)
Source: GTD-CdS database

The propensity to claim attacks—which remained unchanged at the very
beginning of the period—increased sharply after 1975 (Figure 5.5), suggesting that the drive to claim slightly anticipated the increase in the share of
serious attacks. Both are likely to be an effect of the increased group competition among left-wing formations, and reflect the fact that the armed branches
of the major left-wing legal formations, Potere Operaio and Lotta Continua,

Figure 5.5 Relative frequency of claimed acts of political violence by year, Italy, 1974–1980 (N = 1012)
Source: GTD-CdS database

were becoming independent groups and started to claim their acts of violence under new names (Merlo in Cazzullo 2006; Bellosi in Grandi 2005).

The eagerness to claim seems consistent with the idea of political violence as 'propaganda by the deeds', at least as far as left-wing groups are concerned. Groups, for a variety of reasons, seem determined to obtain the credit for their actions. Even the puzzling fact that fifty-three per cent of *failed* attacks were claimed, can arguably be explained by the notion of 'propaganda by the deeds', as perpetrators may still attribute to such attacks a positive communicative value. In 1979 for instance, the BR claimed two failed attacks against army barracks in the centre of Turin. They were experimenting with a rocket launcher, a type of weapon new to their repertoire. They claimed the attacks to mark the threat that the possession of this weapon posed for the state—never mind that they were not yet able to handle it accurately (Peci 1983).

Recurrent and One-off Names

When we look at the *names* used to claim the attacks the plot thickens. The 1012 attacks which were claimed, included 282 names in total (GTD-CdS database): eighty-four of these names (twenty-nine per cent) took credit for more than one action: fifty-nine claimed between two and five actions, thirteen between six and ten, and five between eleven and twenty; only seven names claimed more than twenty attacks over the whole period (Table 5.1).

Table 5.1 The number of names that claimed a given number of attacks, Italy, 1974–1980 (N = 282)

Claiming names	Attacks claimed
7	>20
5	11–20
13	6–10
59	2–5
198	1

Source: GTD-CdS database

The name 'BR' appears in twenty-one per cent of all claimed attacks (215/1012), followed by Prima Linea (PL) with seven per cent (72/1012), and Nuclei Armati Proletari (NAP), with nearly three per cent (30/1012). The only far-right group to claim more than twenty attacks was Nuclei Armati Rivoluzionari (NAR), which claimed just over two per cent (twenty-six). Of these groups only the BR was in existence throughout the whole period 1974–1980; NAP emerged in 1975 and disbanded in 1977; PL claimed for the first time towards the end of 1976 and disbanded formally in 1987; while the NAR had a short and brutish life between 1977 and 1980. These groups, which were the hard-core organizations of the political violence during the period under study, used their actual names when claiming, arguably with the aim to associate their names with the attacks and gain in terms of reputation.

However, the vast majority of names used in claims, 198 (seventy per cent), were used only once and never again. If the purpose was 'propaganda by the deeds' this would be puzzling for one cannot build a reputation, intimidate, and draw resources if one keeps changing its name. So, which are the benefits of claiming once and never again?

The majority of these one-off names had generic left-wing connotations (seventy-one per cent). There was a proliferation of 'fronts', 'nuclei', 'formations', 'brigades', 'squads', and 'commandos'—names used interchangeably by left-wing extremists at the time, who made them up modelling on the vocabulary typical of their camp. Their uniqueness could simply be a by-product of political failure, due either to groups that dissolved in their infancy, or to names so ineptly coined from a 'marketing' point of view, to be still-born (see last section for examples). Overall, one-off names could reflect the large contingent of militants engaged in spontaneous and unorganized political violence (Della Porta and Rossi 1984), as suggested by the fact that most one-off names were used to claim minor attacks.

The only evidence that this phenomenon was more than a by-product of turbulent political times, suggests that some such names were used by the same pool of militants who aimed at mimicking a fighting front bigger than its actual size, to create, as it were, the illusion of a 'prairie fire' (Anonymous 1979; Pisano 1979; Rimanelli 1989). Some extremists got together, engaged in some violent act, and decided to claim it as group x; then for the next attack they chose the name y, and so on. This strategy was theorized by left-wing militants who refused the structured military model adopted by the BR (Boraso 2006). The aim of mimicking a bigger sized group was first to confuse law and order agents, who initially overestimated the capacity of the violent arm of the left-wing movement (Pisano 1979); and next, to embolden unde-cided left-wing militants to join in the violence by making them believe that the uprising was growing fast and wide (Segio 2006).[10]

Claims that used one-off names for those purposes were neither true nor false. The groups claiming with these labels did not have a 'real' name that could or could not be disclosed; at every new claim they just wore a differ-ent 'hat', which they minted and which belonged to no one else. Still, they deliberately kept changing the name at every new attack to send a false signal intended to make people believe that perpetrator x was different from per-petrator y, while making clear that both x and y were broadly fighting for the *same* left-wing cause. They were mimics of size rather than of group identity, unlike those to which we turn now.

False Claims and Mimics

From the GTD-CdS database we could identify 103 attacks that were falsely claimed, eleven per cent of the 1012 claimed attacks. They were identified for a number of reasons: (1) the perpetrators of the attack were caught and unmasked as false claimants; (2) the police expressed serious doubts about the authenticity of the claim and later confirmed to be false; (3) the false claim elicited the reaction of the group victim of this kind of piracy. Clearly this is a lower bound, for false claims that deceived everyone remain unde-tected. This is not so likely for false claims that involved major attacks which attracted greater investigative attention and more robust responses of those groups who were victims (more on this below).

There are two main types of false claims:

---

[10] Tribunale di Genova, Ufficio Istruzione GI Giuseppe Petrillo, Sentenza Ordinanza 11/9/81 vs Baistrocchi Livio+10, Testimonianza di Fabrizio Rainone, September 26, 1980.

(i) Agents who are not the perpetrators claim an attack using their group name. There were twenty false claims of this type; this set is formed by mimics who are in a strict sense *free-riders*, who want to obtain the credit of someone else's 'work'.

(ii) The real perpetrators claim an attack which they carry out, but use the name of a group, real or imaginary, which is not theirs. There were eighty-three false claims of this type; mimics are false-flaggers, but their aims vary: they include blame-shifting, whether for defensive or aggressive reasons, and some free-riding too, in the sense that carrying out an action using a name with a fierce reputation may make the action more intimidating.

In (i) it is the attack to be falsely claimed: in (ii) it is the name to be false.

The information surrounding six of the 103 false claims is too muddled to establish with any certainty the identity of the mimics and their purposes. If we slice the remaining ninety-seven false claims by their purpose, we see that forty-five were issued by criminals, while fifty-two were politically motivated (Table 5.2).

The former were issued mostly in connection with extortion, kidnappings, and robberies, while the latter occur mostly in connection with violence against people, such as assassinations and bombings (Figure 5.6).

If we cross the types of false claim with the identity of the mimics, we see that type and identity map out neatly: type (i) false claims were mostly carried out for political reasons by left-wing groups who claimed responsibility for actions committed by some other left-wing group (Table 5.3).

Free-riders, as Figure 5.7 illustrates, mostly appear in connection with two types of acts of political violence: assassinations and wounding which were the targeted violence typical of left-wing extremists.

False-flaggers instead are spread over a wider spectrum of acts of violence, most of which of a kind also perpetrated by ordinary criminals. There were

Table 5.2 Number of false claims by types of mimic and types of perpetrator, Italy 1974–1980 (N = 103)

	Criminal	Political	Unknown	Total
'false flaggers' (Type i)	44	33	6	83
'free-riders'(Type ii)	1	19	0	20
Total	45	52	6	103

Source: GTD-CdS database

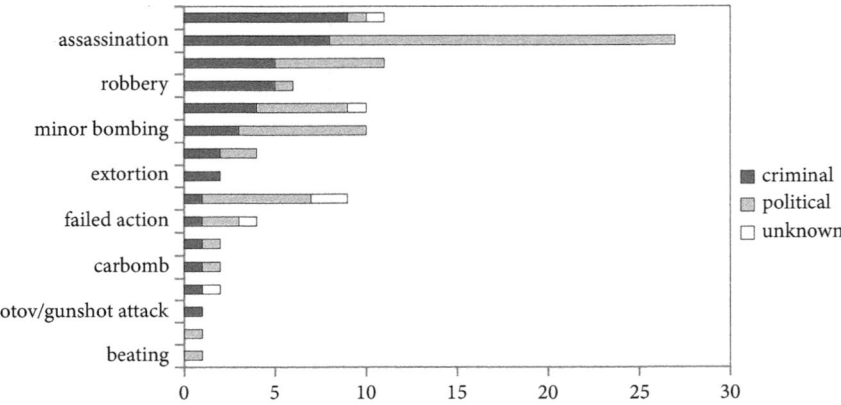

Figure 5.6 Number of false claims by types of attack and types of perpetrator, Italy 1974–1980 (N = 103)

Source: GTD-CdS database

Table 5.3 False claims by types and by the identities of the mimic (N = 103, 6 missing values)

Type of false claims	Identity of the mimics Political Left-wing	Right-wing	Criminal	Total
(i) 'free-riders'	17	2	1	20
(ii) 'false flaggers'	14	19	44	77
Total	31	21	45	97

Source: GTD-CdS

two main purposes for false claims of type (ii): some were made by right-wing groups, who carried out an attack and tried to put the blame on left-wing militants; others were issued by criminals who exploited the names of political groups either to increase their intimidation power or to confuse the investigations or both. We shall now review their actions in more detail.

Criminal Mimics

Conflict situations offer special opportunities to criminals by extending the pool of violent political groups' names in circulation that can be used fraudulently. During a robbery or a kidnapping, or while attempting extortion

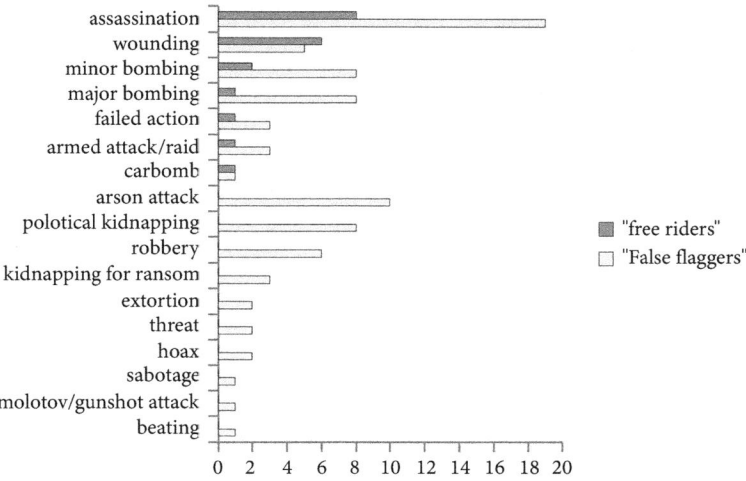

Figure 5.7 Number of false claims by types of attack and by types of mimic, Italy 1974–1980 (N = 103)

Source: GTD-CdS database

(see Mamidi, this volume), criminals would drop in the name of a menacing political group to scare the victims into compliance; or they would claim under the name of a political group to put the police on the wrong tracks. In both cases criminals are the real perpetrators who use the name of some existing or invented political group.

In the first case, in order to enhance their chances of success, criminals exploit the capital of intimidation contained in the name of the group they use. There were amateurs, like two enterprising schoolgirls who issued extortion letters signed with a fanciful 'Black Brigades' (CdS 21.1.1975), and a Romanian exile who pretended to be the leader of the BR in a futile attempt to cash in on Moro's kidnapping (CdS 16.3.1978). But criminals were not always so incompetent: a lone bank robber pulled off a big heist in Genoa by presenting himself as a member of the BR and threatening the group's retribution against uncooperative staff's families (CdS 12.4.1978). A resounding name is more likely to be known by the victims of a crime and more effective at 'beating' them into submission without the use of force: in fact, rather than making up names, in fourteen out of fifteen false claims aimed at intimidation the criminals issued the names of well-known groups (seven used the name 'BR').

When not intimidating their victims, criminal mimics aimed at confusing investigations. In 1979, for instance, the assassination of Judge Cesare Terranova in Palermo, Sicily was claimed under the name of 'Ordine Nuovo', a far-right formation (CdS 25.9.1979), and was issued to distract attention from the Sicilian Mafia, which was the real culprit. The actions that were the object of a criminal false claim received for the vast majority only that one claim (thirty six of the 45); and even when there appeared to be multiple claims, in reality they were likely issued all by the same mimic to confuse investigations. After the assassination in 1980 of Piersanti Mattarella—the president of the Sicilian Regional Assembly and head of the Christian Democratic Party in Sicily—the Sicilian Mafia claimed it under the name of a non-existent 'Nucleo Fascista Rivoluzionario', and then, for political 'equanimity', they issued further claims as 'BR' and 'Prima Linea' (CdS 1.7.1980), neither of which had ever carried out attacks in Sicily where they had no political base. As a result, the assassination was initially deemed as the Sicilian counterpart of the assassination of Aldo Moro when in fact the murderous hand was again the Sicilian Mafia.

Extortion and kidnapping for ransom seem to attract the highest share of false claims relative to the total claims of the corresponding type (Figure 5.8).

But this does not mean that these criminal acts were the most inspiring for false claimants. Criminal acts enter in the database only if their perpetrators issued a claim using a political name. Thus, paradoxically, the only ones who claimed such acts using a political name, and ended up in our database, were the real criminals who rather than just remaining anonymous chose to go

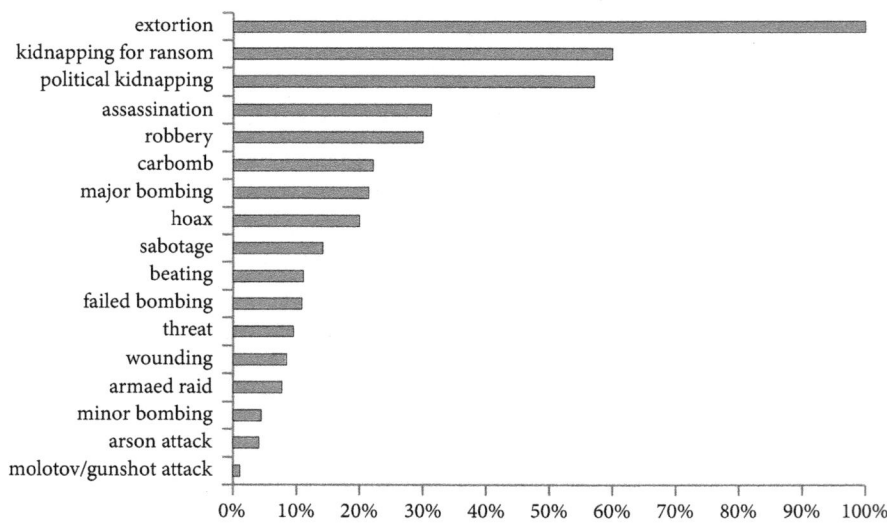

Figure 5.8 Relative frequency of false claims over total claimed attacks by types of attack, Italy, 1974–1980 (N = 1012)

Source: GTD-CdS database

one step further and pass themselves off as political agents. Purely criminal acts carried out by purely criminal agents which went unclaimed are not in the database, and therefore the numerator is close to the denominator, which makes it seem as if these types of action attracted a very high share of false claims.

Ironically, *political* groups who truly perpetrated such acts in order not to be discredited often preferred to remain anonymous or even to mimic being ordinary criminals!. Robberies and kidnappings, which were among the repertoire of self-financing actions (Marighella 1969, 6–7, 13, 42, 52), caused much unease amongst left-wing militants who dreaded being confused with villains (Paroli in Bianconi 2003, 18–19; Franceschini, Buffa, and Giustolisi 1988; Bellosi Grandi 2005, 44). Mario Moretti, a leader of the BR, explained why during the 1970s they kept them anonymous:

> How does a robbery in the name of the proletariat ('*esproprio proletario*') differ from a pure and simple robbery? The only difference is that the perpetrator should be recognized by the movement, and should have credibility as part of a revolutionary organization. We could not take this for granted as we were yet to prove it [in 1972]. We had to wait until '77, when we kidnapped [Pietro] Costa, the ship-owner, to claim an '*esproprio*' as BR (Moretti 2000, 23 our translation).

Right-wing groups, too, hid their political identity when committing crimes; for instance, when engaged in robberies in a mixed gender crew, they tried to pass the women off as men, since women, absent in teams of ordinary criminals, were a give-away of politically motivated actions; or when they chose to engage in criminal behaviour which was routinely attributed to drug addicts—such as stealing, during a bank robbery, the customers' gold chains and bracelets (Fioravanti in Bianconi 2005, 120, 225).

Political Mimics

The fifty-two false claims that were politically motivated are of two main kinds. The first is made mostly by right-wing provocateurs who sign an act of violence they commit with the name of a left-wing group—a classic false-flag operation. This kind of mimicry occurs in nineteen cases out of the fifty-two (see Table 5.3). The main aim was fuelling the fear of the 'red danger'.

Right-wing militants issued false claims for other reasons too. In one instance they tried to cover the murder of one of their militants, Martino Traversa—suspected of being a police informant—claiming it under a fictitious left-wing name (CdS 13.3.1980). On other occasions, right-wing militants claimed robberies using left-wing sounding names (CdS 6.3.1980). The primary motives of these acts were punishment and self-finance and could have been left anonymous. Mimicry offered the added bonus of shifting the blame onto the enemy.

When carried out to harm a particular group, false-flagging by definition must use the name of *that* group. Also, as we have seen, when the purpose is intimidation, fictitious names are not helpful, as only recurrent signatures serve this purpose as they embody the threatening reputation of known groups. When instead, false-flagging simply aims to confuse the investigations and shift the blame for defensive purposes, mimics can also gain by using invented names. These have the advantage that they cannot be exposed as false because there is no established template of a recurrent name to compare them with, and do not prompt the reaction of an existing group. (A compromise was to use names of existing *foreign* groups, like 'Montoneros', an Argentine left-wing urban guerrilla group that was flagged twice probably by right-wing mimics).

In some cases left-wing groups, too, used false flagging (see Table 5.3). This ruse, for instance, was used by militants of a particularly wayward left-wing group, Autonomia Operaia, who kidnapped a child, Guido Freddi, and asked for a ransom using the name BR (CDS 20.8.1979). The BR were not amused at being blamed for such a vile act and responded forcefully as we shall see below. In another case, a group of workers sabotaged the plant equipment at Fiat near Turin, and claimed it as the BR, and again the BR disowned the action because they were reorienting their strategy outside the factories. In a third case, militants of PL blamed the botched operation on which two policemen lost their lives on the BR.

However, the majority of left-wing mimics is made up of those who try to take the credit for acts they did not commit. This kind of mimicry occurs in nineteen cases out of fifty-two (see Table 5.3). (In the database, these cases appear as attacks which attracted more than one claim of responsibility, in which usually the first one was true and the subsequent ones false). In these false claims the mimics were other left-wings formations of lesser standing than the BR (we found no evidence of right-wing groups free-riding on each other's actions). For instance, in twelve out of fifteen multiple claimed attacks in which the first is by the BR, the subsequent competing claims were issued by another left-wing group. Proletari Armati Per il Comunismo (PAC)—a left-wing formation active between 1977 and 1979—added their claim to the wounding of a journalist which had been carried out by the BR (CdS, 24.4.1979). PL—the second strongest violent left-wing formation competing for supremacy with the BR—appended their name a few times to actions carried out by the BR, and at least once the BR 'reciprocated' and did the same to PL.

The best outcome for a free-rider is to be publicly recognized as the real perpetrator of the attack. This outcome, however, was unlikely to be achieved as honest perpetrators not only issued convincing counterclaims, but soon learned, as we shall see below, to release claims in the immediate aftermath of the attack to pre-empt free-riding mimics. The second best outcome was to achieve some publicity, have their name appear even fleetingly next to the name of a stronger and more established group—photo bombing, as it were, in the other group's moment of fame. The target audience of the stunt was not the public at large, but presumably the left-wing constituency that the mimics shared with the BR.

Overall, for violent attacks which were politically rather than criminally motivated, the frequency of false claims by type of action claimed is more reliable (Figure 5.8): the highest shares of false claims over the total of claimed attacks of each type belong to political kidnappings, assassinations, and major bombings. It makes intuitive sense that if one is into issuing false claims, there is more to be gained by claiming serious rather than minor attacks. However, we cannot be sure that our distribution fully reflects the false claim propensity. The distribution of false claims could be the outcome of the relatively stronger effort made by law and order agents trying to identify those responsible for major acts of political violence; or of the reaction of the groups who suffer from that, either because they are falsely accused of an attack and unwilling to take the blame, or because they are enraged at seeing someone else take the credit for what they did. It is to consider these responses that we turn next.

Models' Response

What about the groups that were targeted by mimics? Some of these saw their names attached to actions they did not do; and others, who carried out a violent act for which they sought credit, saw free-riding 'names' trying to snatch it away. How did they defend themselves?

There are two main ways in which one can defend one's group name: preventing mimics and exposing them ex-post. As a last resort, there is, in principle, a third option: changing a name when it becomes tarnished or out of control. This option, however, is unlikely to succeed. The story of Nuclei Armati Rivoluzionari (NAR) is exemplary. In 1979, Valerio Fioravanti, leader and founder of this far right-wing group, had lost control over the name:

'everybody—he said later—had taken possession of the acronym and it had been involved in all sorts of actions' (Bianconi 2005, 182). He decided to rename the group. In order to publicize the move with an announcement in which NAR was declared defunct and a new signature introduced, he organized a major attack against a police patrol and the assassination of a judge. Switching the old name off, however, proved arduous. The old name still had to be mentioned if only to declare it deceased and as a result it kept hovering over the group: 'formerly known as NAR' or 'New NAR', as members themselves began to call the group (Bianconi 2005, 182–186, 199–203, 233). Given that disassociating from a corrupt name, retaining one's reputational capital, and repackaging it into a new credible name can prove very difficult indeed, a group has a strong incentive to defend its name by other means (Gambetta 1993, 144).

Exposing Mimics

Ex-post, a group can expose mimics either by a disclaimer or by a counter-claim, depending on the type of mimic attack they suffer. A disclaimer is a statement by which a group denies their involvement in an attack that has been claimed under their name. A counterclaim is a statement by which the perpetrating group challenges a free-rider's false claim.

Some of the names used in false claims did not cause harm to a particular group. These were one-off names—twenty out of ninety-seven false claims—generically modelled on the left-wing camp vocabulary either by criminals or by right-wing provocateurs. None of these false claims prompted a reaction, as they did not hurt a specific group capable of organizing a response—with one exception: when right-wing militants blamed the assassination of Martino Traversa (see earlier in this chapter) on a fictitious name which implicated the left-wing movement as a whole, activists managed to take collective action and issued a disclaimer.

The other seventy-seven false claims targeted recurrent names, which belonged to groups determined to build a reputation and maintain control over the public understanding of their aims and violent acts. The twenty claims issued by free-riders—seventeen by left-wing mimics, two by right-wing, and one by a mythomaniac—attracted four counterclaims, two released by the BR, and one each by PL and the NAR. The false claims issued by a variety of false-flaggers and with the potential to hurt a group by attributing to it an action they did not commit, were fifty-seven, and attracted fifteen disclaimers by established groups: six by the BR, three by the NAR, two by PL, and one each by Nuclei Armati Proletari, Ronde Proletarie, Ordine Nuovo, and Ordine Nero.

One may wonder why there were relatively so few counterclaims and disclaimers. One reason is that groups did not always bother to disclaim because they did not need to: the mimicry attempt failed either because the police caught the mimics or because the false claim was shoddily enacted and everyone could see it was a fake (itself often a result of preventive signature design, see next section).

A second reason is that constructing a convincing disclaimer is hard; disclaimers were often met with scepticism and suspected of being 'crocodile tears'—not without motive:[11] three disclaimers of fifteen were themselves *false*, attempts to shift the blame of actions that had gone badly wrong onto innocent groups.[12] The ineffectiveness of disclaimers, however, does not imply ineffectiveness of counterclaims: if true, it is easier to muster evidence of a group's culpability than evidence of a group's innocence. Something else must explain the dearth of counterclaims.

A third reason is that both disclaimers and counterclaims were only issued in cases in which the act of violence being falsely blamed on an innocent group or falsely claimed by mimics was very serious. Figure 5.9 shows the numbers of disclaimers and counterclaims and compares them with that of falsely claimed acts of violence.

The ranking seems consistent with the choices of actors who care about their political reputation and the consequences of their actions. It is worthwhile to issue a disclaimer to remove responsibility for acts of indiscriminate violence (major bombing) and major crimes such as assassinations. Minor attacks instead (beatings, arson, and Molotov/gunshot attacks) elicit no disclaimer. It seems that disclaimers were also worthwhile in order to maintain political control over actions that could have fallen into the domain of

---

[11] NAR's disclaimer of the bombing at Bologna train station in 1980 was not believed (CdS 6.8.1980), although there are now doubts about the involvement of the group in that atrocity.

[12] In 1979 the BR made a bad mistake. An action intended to maim, resulted in the victim's death—a Communist trade unionist in Genoa. After two days they issued a communiqué denying responsibility, but two weeks later, in another communiqué, claimed responsibility (CdS 24.1.1979; 9.2.1979). In 1980 Ronde Proletarie (a splinter group of Prima Linea) killed a private guard. The action was meant as a testing exercise for new recruits and the aim was taking the guard's gun and not his life. Soon after a phone call claimed the wounding of the guard who had died in the meantime. A disclaimer was issued once Ronde Proletarie realized their mistake. The authors, young Prima Linea recruits confessed it later, after they were apprehended (CdS 10.4.1980; Novelli and Tranfaglia, 1988). In 1978 a bomb at the *Gazzettino Daily* in Venice caused one death. A phone call soon after claimed it as Ordine Nuovo, then another phone call in the evening claimed that the authors were the BR, and a day later Ordine Nuovo issued a communiqué explaining that they did not mean to cause victims and wished they never carried out the attack (CdS 22.2.1978).

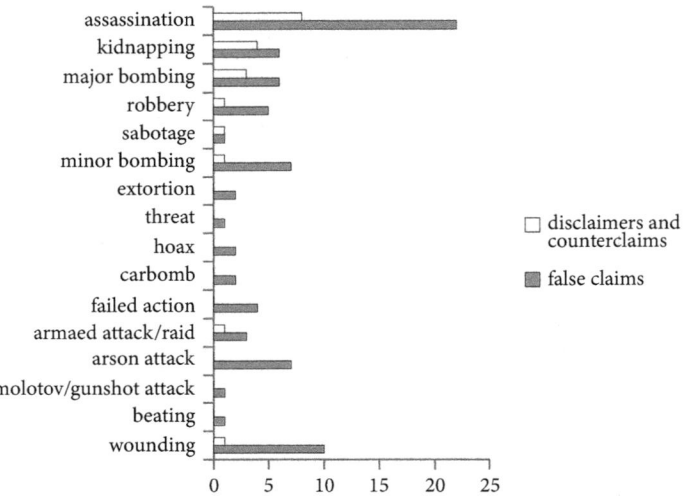

Figure 5.9 Numbers of disclaimed and falsely claimed attacks by types of attack, Italy 1974–1980 (N = 19/81) [one-off excluded]
Source: GTD-CdS database

ordinary criminals (kidnapping, robberies, and minor bombings). Minor incidents seemed not worth bothering with.

The BR used its six disclaimers when the attack they would have been saddled with had heavy reputational costs. After a botched robbery in which three policemen were killed, for instance, the BR issued the following statement:

> Let Autonomia Operaia or Potere Operaio or Bande Armate, or whoever they are, know that they must claim with their own acronyms. Should they use our name again we will turn them in. We will deal with them as we dealt with the members of Autonomia Operaia who kidnapped young Freddi' (CdS 25.3.1980).

This was a credible threat, as what eventually cleared the BR of responsibility over Guido Freddi's kidnapping was not the disclaimer that they duly issued at the time, but their tip-off to the police leading to the discovery of the typewriter of the militant of Autonomia Operaia who had produced the false claim (CdS 20.8.1979). The threat was credible, but, as it turned out, they were barking at the wrong mimic: the authors of the botched robbery were not, as the BR suspected, their left-wing competitors, but ordinary criminals!.

In conclusion, disclaimers were issued sparingly when it was possible to prove one's innocence or responsibility, and as a measure of last resort when the feared consequences of the false attribution or the snatching of due credit could have been serious, even though it might not sort out the desired effects. As a tool to enforce property rights over their name, however, the outcome of a disclaimer is quite uncertain and the cost of producing a convincing one can be too high even for the owner of the falsely issued name.

Preventing Mimics

Being able to prevent mimicry is preferable as it reduces the cost of monitoring and taking remedial actions in order to enforce the property rights over a group's name or attack.

The technique of prevention depends on the type of mimics. A way to prevent free-riders, who mimic an attack and claim it with their own name, is for the real perpetrators to carry out their actions with a distinctive pattern and repeat it in every subsequent action; the best way, however, is to claim responsibility giving at the same time evidence that only the perpetrators could possess, and no one else can. A way to prevent false flaggers, who mimic the name and shift the blame of an attack on an innocent group, is instead to design a 'signature' which relies on more than a mere 'disembodied' name.

Let's consider name protection first. Think of a claim made by phone. The name of an organization whispered down the receiver may seem the cheapest of all possible words, which anyone could utter. Yet, the voice, the accent, the issuer's speech idiosyncrasies, and the manner of the claim delivery, once linked to an action and repeated in exactly the same details over new actions, can establish a name and its credibility, and, taken together, become hard to imitate (Bacharach and Gambetta 2001). Code words issued in connection with a name are another way to protect the name (Hamill, this volume), however, there is no evidence that this system was used in Italy. In case claims are made in writing, one can give one's name some degree of mimic-resilience by executing the name and the logo in a particular style, using a particular pen, ink, paper, or printer uniquely identifiable by some feature of theirs, including systematic errors. A unique combination of oral or written elements can become a *signature*, which makes the identity of the caller recognizable from one instance to the next as the same signature. It becomes like a fingerprint that reidentifies a claimant as the one who 'honestly' claimed before and supports the honesty of the present claim. In short, making a signature mimic-proof is the result of a set of elements that raise the cost of producing it to a point at which only the group that rightfully owns the signature can afford it, and false claimants find it too costly. In the next section we review evidence that the BR considered these features.

Consider now protection from free-riders claims. The type of targets, the choice of weapon, the style of execution, and of claiming methods can, if repeated, go some way towards making it hard to claim an attack which one has not truly carried out. But the royal technique to stop free-riders from intruding is to claim an attack linking the claim to evidence of a kind that no one else other than the perpetrator can possess. Table 5.4 compares how true and false perpetrators delivered their claims.

Most claims were delivered by phone to newspapers or news agencies (the Internet was a long time to come). It seems that genuine claimants can afford methods that are costly for false claimants, if not outright impossible. The latter deliver their claims mostly through simple phone calls or

Table 5.4 Claiming methods by true and false perpetrators

Claiming methods	False claims (N = 94)[a]	Right-wing claims (N = 93)	Left-wing claims (no BR) (N = 506)	Red Brigade (N = 158)
Cheap				
Phone calls	56%	28%	27%	9%
Hand-written document	6%	12%	1%	1%
Total	62%	40%	28%	10%
Some cost				
Simultaneous phone calls	2%	2%	3%	7%
Phone call and printed document	14%	23%	26%	35%
Total	16%	25%	29%	42%
High cost for 'free-riders' (but not for 'false-flaggers')				
During action	16%	31%	17%	9%
During action followed by document	0	0	6%	8%
Communication of exclusive clue	2%	1%	5%	9%
Phone call before action is made public	4%	4%	14%	23%
Total	22%	36%	42%	49%
Missing data	9	7	74	14

Source: GTD-CdS database
[a]Here we only count the claiming methods of the second claiming group. The sources give details about them only for a reduced number of cases.

hand-written documents (sixty-two per cent). Relative to other claimants, left-wing ones show a much larger preference for methods that have at least 'some cost' of production (e.g. synchronized phone calls, phone call followed by a printed document). The BR used high-cost claiming methods more than anyone else (forty-nine per cent). Exclusive knowledge was displayed in claims issued immediately after an attack, before radio and television news bulletins would broadcast it.[13] Once the event was known, exclusive details—about the action, the weapons, and even accidental mistakes—were revealed as proof of authorship. Phone calls used by the BR were never a simple announcement. They were laced with exclusive knowledge and contained instructions on where to find communiqués dense with ideological statements as well as details about the target. These details stood as a barrier against free-riding mimics who could not afford that kind of knowledge about the execution of the action or the ground-work carried out in order to accomplish it.

Note that exclusive knowledge is of little use when the perpetrator is not a free-rider but a false-flagger. Table 5.5 shows the methods of claiming by type of mimics. Here we notice that the claiming behaviour of false-flaggers and free-riders differs in a significant fashion.

Free-riders, in our case at least, use *only* phone calls. False-flaggers instead use a wider range of methods, including what for free-riders would have an impossibly high cost, such as claiming during an action or giving exclusive details of the action: they know full well the details of an attack for the very reason that they are the real perpetrator.[14]

Table 5.4 also reveals that right-wing true claims have a pattern that differs from that of left-wing true claims. The former were not targeted by agent provocateurs, and we can surmise that they can thus 'afford' cheap claiming methods. There is, in fact, a strong correlation in terms of claiming method between true right-wing claims and false claims ($r = 0.70$), while it is weak between true left-wing claims and false claims ($r = 0.10$).

---

[13] Unlike the IRA who issued advance warnings, the Italians were more prudent. The BR, for instance, had a rule never to write the document explaining an act of violence before the completion of the mission so as to avoid incrimination in case the action was foiled (Moretti 2000: 137–138; Guagliardo in Rossa and Fasanella 2006: 24).

[14] This occurred in two assassinations—the industrialist Attilio Dutto in Cuneo in 1979, claimed as the BR before the news had reached ANSA (CdS 21.3.1979) and Michele Reina, a Christian Democratic politician in Palermo assassinated in 1979 was claimed as PL soon after the event (CdS 13.3.1979)—and in the wounding in 1978 of Girolamo Mechelli, a Christian Democratic politician in Rome which was claimed as the BR soon after the event (CdS 26.4.1978). The killers of Agatino Coniglione, a petty criminal and police informer, signed as NAP and to prove they were the authors they announced the name of the victim to the police before it became public (CdS 29.7.1977). None of these crimes were political, and the culprits, with the exception of Dutto's murder, have been identified.

Table 5.5 Claiming methods by type and purpose of false claims

	'False flaggers' (N = 75)	'Free-riders' (N = 19)
Cheap		
Phone calls	52% (39)	74% (14)
Hand-written documents	7% (5)	4% (1)
Total	59%	76%
Some cost		
Simultaneous phone calls	1% (1)	0
Phone call and printed document	12% (9)	22% (4)
Total	13%	24%
High cost for 'free-rider' (but not for 'false-flaggers')		
During action	20% (15)	0
Communication of exclusive clue	3% (2)	0
Phone call before action is made public	5% (4)	0
Total	28%	0

Source: GTD-CdS database

On balance it seems that left-wing groups found it harder to protect themselves against false-flaggers than against free-riders, and had to make a significant investment in designing a mimic-proof signature. This is what the BR did: because of their success and reputation they became the main favourite target of mimics. The name BR was used in thirty-five out of the 103 false claims. Twenty-nine of the thirty-five false claims occurred between 1977 and 1980 when the group faced both more competition and reached the apex of their political impact. They attracted mimics of both types—twenty-three were false-flaggers and twelve were free-riders—and we shall examine how the BR reacted to the threat posed by both in the following section.

The Making of a Signature

Within the left-wing movement there were two conflicting views on who should be fostering the revolution: Lotta Continua, Potere Operaio, and Autonomia Operaia identified the diffuse left-wing movement as the spontaneous engine of revolutionary impulse, while the BR took a 'Leninist' approach and identified it with a secretive underground vanguard who by targeting the elites and inducing brutal state repression would awaken the masses to take up arms (Della Porta 1995; Moss 1989). When the BR started claiming their violent actions they caused division within the movement. According to the leaders of Lotta Continua:

> we were disturbed when the BR started to sign their actions: we feared they would carve out their own political space (...) We wanted to keep everything [including violent actions] under the control of the [legal] organization: whoever was going beyond it, was considered a rival because they were enclosing a space which we wanted to keep connected in order to prevent the fragmentation of the movement (Merlo in Cazzullo 2006, 193 our translation).

The armed branches of the legal groups would strike anonymously against targets clearly belonging to the opposite camp, thus gaining the credit of the attack for the movement as a whole, with no need to sign with a name. The BR's decision to sign their actions was perceived as an attempt to turn this collective capital of political violence into a 'private good' (Enders and Sandler 2006, 37). The BR thought otherwise: 'We are the Red Brigades not just one group among the others' (Moretti 2000, 220, our translation). They wanted to keep control over the strategy of violence, and were opposed to letting spontaneity have a free hand. Furthermore, they wanted to radicalize the struggle and signing was seen as a way to recruit those truly ready to embrace the violent route.

But signing is neither simple nor without cost. It was the spring that made communication strategies salient in militants' reasoning. Finding a name was, of course, the first requirement. They minted 'Brigate Rosse', inspired by Brigate Garibaldi, the Communist branch of the Italian Resistance in the Second World War (Podda 2007, 90); they replaced 'Garibaldi' with 'Red' in order to remove the reference to the 'bourgeois hero' of the Risorgimento (Franceschini, Buffa, and Giustolisi 1988, 32).[15] In naming their group they acted as advertisers branding a product: the name should be easy to remember, distinctive enough not to be confused with any other in the same domain, and somehow evoke the cause the group is fighting for.[16]

Not everyone was so clear-headed about the properties that an effective name should have. In 1978, an arson attack near Varese was signed by the 'Primo Reparto Comunista Combattente del Fronte Operazioni Studi

---

[15] One wonders whether the BR also conceived the name in opposition to 'Brigate Nere' the Fascist paramilitary groups operating in northern Italy for the Repubblica Sociale during the final years of the Second World War.

[16] 'Brigate Rosse' was not their first choice, they preferred 'Lotta Continua', building on the French 1968 motto 'Continuons le combat!' which was already taken (Boato in Podda 2007: 64).

Informatica Militare', which unsurprisingly remained a one-off name (CdS 29.5.1978); in 1979, a Nucleo Armato Rivoluzionario Nazionale signed the bombing of a FIAT outlet in Milan and from this unusual mix—revolutionary versus national—the receivers of the claiming call were left wondering whether it was a left- or a right-wing claim (CdS 5.11.1979); in 1978, the letters RCC were scribbled on a wall to sign an armed attack, but nobody could work out its meaning until the Reparti Comunisti Combattenti called in to reveal the acronym (CdS 29.6.1978). In late October 1977 there was a wave of attacks throughout Italy in reaction to the suicides of the members of the Rote Armee Fraktion in Stammheim prison: several times the perpetrators had to call back to have their name rectified as they were too abstruse for the newspapers phone operators to recall them accurately—mishaps reminiscent of a famous episode of Monthy Python's *Life of Brian,* which came out in 1979.

After the name, the BR needed a logo. Their five-pointed star first appeared at the end of 1970 in Milan, when the group was still semi-clandestine and counted only a handful of members (Figure 5.10). Its choice came as a result of an elaborate reflection:

> We had been thinking for a while about a symbol to represent us, BR's founder Alberto Franceschini later explained; it had to be easy to grasp and simple, easy to draw, even on walls. We were in no doubt that it had to be a star. It was the flag of all revolutionary armies: Vietcong, Tupamaros, Che Guevara, Brigate Garibaldi. However it had to be recognizable as our own, *we had to make it exclusively ours.* We thought that a hundred lire coin that any comrade could carry in his pockets was a good device to trace the circle containing our star. We tried to draw it with a technique we learned at school […] without lifting the pencil from the paper. We could never draw a flawless star […] and we decided *which mistake* all of us had to make: a shorter upper tip to give the impression that the star aimed high into the future (Franceschini, Buffa, and Giustolisi 1988, 62, our translation and italics).

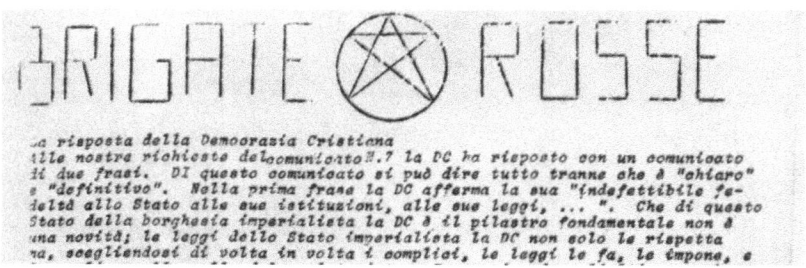

Figure 5.10 Red Brigade's logo

In the first half of 1971, there was a proliferation of minor actions signed 'BR', most of which perpetrated by out-group supporters. They were spontaneous acts hitting targets favoured by the BR—cars belonging to company executives, factory managers, and right-wing trade unionists—and used the same kind of cheap weapons—incendiary devices. Far from damaging their model, these friendly eager mimics provided an inflated perception of its might. Not only did the BR not complain, at first they welcomed having their insignia used widely.

However, they soon realized that the simple combination of name and logo could work as easily for as against them. The need to increase their control over their signature came in 1971, when a series of bombing attacks against factories and army barracks were falsely signed as BR. The explosive used in these attacks was the same as that used in an indiscriminate bombing campaign underway at the time, and generally attributed to far-right militants. This reinforced the suspicion that 'BR' was a name used by right-wing agents provocateurs and not a name of a real left-wing organization. The BR's reaction was swift and multi-pronged.[17] First, a communiqué denounced the claim as a plot of right-wing militants and the police. Second, to prove their left-wing credentials they endorsed a series of arson attacks that left-wing militants had carried out against right-wing targets in Rome. The unaffiliated militants had signed these attacks as BR, and the BR ex-post laundered them and turned them to their advantage. Attacks on right-wing targets remained a strong signal of bona fide left-wing credentials whenever a new BR column was formed (Bianconi 2003, 29; Moretti 2000, 74; Pisano 1979, 52). Finally, the BR introduced a strong constraint on their choice of weapons which pulled the rug from under the false-flagger's feet: bombings were ruled out of their repertoire, and this was advertised so as to avoid being blamed for them; indiscriminate violence was stigmatized as the weapon of right-wing and police agent provocateurs.

The BR's signature developed over time to meet the threat of free-riding mimics too. It included a repetitive style of execution[18] and the use of weapons as 'fingerprints' (Fiore in Grandi 2007, 98; Peci 1983, 14; 'Paolo' in Raufer 1993, 318). In several knee-capping and executions, the BR used a Nagant 7.62, a Russian revolver built in 1895, and last produced in 1944. Since this revolver is laborious to reload—the cylinder does not swing out of the frame

---

[17] For a full description see www.bibliotecamarxista.org/soccorso%20rosso/capitolo%206.htm.
[18] BR members were so methodical in executing their actions, among which knee-capping was frequent, that they were mockingly nicknamed, in Piedmontese dialect, 'travet 'd la tibia'(the petty clerks of the shin). Ignazio Travet was the protagonist of a play, 'Le miserie 'd Monsu Travet' written in Piedmontese dialect in 1863 by Vittorio Bersezio. The character was a sad, unimaginative clerk devoted to precision who came to represent the misery of a category of people. The nickname was minted by Enrico Deaglio, a prominent member of Lotta Continua (CdS 24.5.1999).

and one has to remove and insert cartridges one at a time—one can conjecture that it was probably acquired at first because it was cheap and easier to find, but then continued to be used despite better alternatives to leave an identifiable mark.[19] The composition of the hitsquad compensated for the pistol inefficiency as it included one man in charge of the first hit and two more in charge of correcting mistakes or weapon malfunction (Peci in Moran 1986, 36–47).

A great deal of thought went into *how* to issue claims. At first, they kept it simple, scribbling on factory walls or dropping leaflets in factories locker rooms and toilets (Franceschini, Buffa, and Giustolisi 1988, 41; Tessandori 2004): 'Throughout Italy the BR set on fire about a thousand [cars], even though the newspapers hardly ever reported them [...] we didn't care [...] as long as the factory workers kept talking about them' (Fiore in Grandi 2007, 48; Peci 1983, 146–147).

Their signature was fortified with more elements of authenticity when their actions became more daring, and went beyond the factory gates. On 25 January 1971 the first operation that gained them notoriety was a series of coordinated arson attacks against the Pirelli factory in Milan; they signed it by a phone call to local newspapers promising an explanatory document later.[20] Their first kidnapping, on 25 February 1972, lasted just the time to pack Idalgo Macchiarini, a Sit-Siemens executive, into a van, take his picture with the five-pointed star in the background, and a gun pointed to his head. After driving around for half an hour, they abandoned the van and its bewildered content next to a communiqué signed 'Brigate Rosse'. Ironically, local newspapers were on strike and the action failed to gain notoriety. But they did not give up and on 8 March sent a parcel to the Italian news agency ANSA containing the picture of the kidnapped executive in captivity, a copy of the communiqué, and a short message restating their claim, which through ANSA reached every national newspaper (Tessandori 2004, 76–79).

It became standard practice for the BR to call two newspapers—one local and one national—and ANSA to advertise their claims. Interestingly, the credibility of their claims was further strengthened by the decision to claim also their botched operations:[21] 'Whatever we do, cock-ups included, we sign it, as long as we did it' (Moretti 2000, 220 our translation)—a ruse that had

---

[19] www.exordinanza.net/schede/nagant.htm; CdS, 12.3.1978.

[20] Ironically, the Italian Communist Party and Lotta Continua suspected that this major arson attack signed by the BR was a false flag Fascist provocation (Ruggiero 2007, 91).

[21] In June 1974 the newly formed BR column in Veneto botched its inaugural action, accidentally murdering two Fascist militants during a raid of the right-wing party Movimento Sociale Italiano (MSI) branch in Padua. For forty-eight hours they debated on what to do, then a communiqué was issued claiming it and explaining the event as a mistake (Moretti 2000; Bianconi 2003).

a classic predecessor, Cicero, who, facing death, chose not destroy any of his letters, including those which showed him in a bad light, to lend authenticity to his archive.[22]

The signature became particularly elaborate during the kidnapping of Aldo Moro, and with good reason. In the couple of days following the kidnapping, the telephone operator of the charity that volunteered to collect the BR demands, received at least twenty false claims (CdS 19.3.1978, 24.4.1978). The manner of claiming, however, certified the authenticity of the BR's initial claim: the action took place in Rome at 9:15 a.m. and the first claim was issued by the BR simultaneously to local newspapers in four different cities barely an hour after the event, at 10:15a.m., promising to be followed by a communiqué.

They were also careful to 'fingerprint' all communications during the subsequent negotiations using, among others, a technique they employed to respond to a previous episode. When they kidnapped Judge Mario Sossi in Genoa in 1974, there surfaced a forged communiqué. The BR immediately disclaimed it, announcing that they would certify the authenticity of all subsequent communications by using the same typewriter, an Olivetti Lettera 22 (Fasanella and Franceschini 2004, 137). Typewriters leave tiny but uniquely identifying features in the way they execute their characters, and careful examination can establish whether documents have been typed by the same machine.

About Moro's negotiations, Mario Moretti, one of the BR leaders, explained that in all nine communiqués they issued 'the same [IBM] printing head, always the same characters, and always the same way to spread the nine communiqués': each message was delivered the same day at the same time to ANSA and six different newspapers, in the four major cities where the BR had established a column (Milan, Turin, Genoa, Rome). 'It became our trademark', —Moretti continues, 'it conferred authenticity to every communiqué, and made them impossible to counterfeit'. In addition, the BR displayed a combination of cues that no one else could afford: exclusive knowledge, complex communiqués, pictures, coordinated messages. The aim was also 'to show our might by involving the whole structure of the organization, no comrade excluded' (Moretti 2000, 139). Their effort was successful. When someone released a fake communiqué, its features compared with the genuine ones associated to the previous message easily revealed that it was a forgery: the typewriter was similar but not the same as the others, the phone

---

[22] This is according to Robert Harris's biographical novel *Dictator* (2015, 487) on Cicero's later years. The great Roman orator left seventy-two volumes of correspondence to posterity, although thirty-five of them are now lost.

call was made by a voice with a strong Roman accent rather than a northern Italian accent as previously, the name BR was drawn differently from the standard, and the pattern of distribution was anomalous (it was found in Rome and Ancona rather than Rome, Milan, Genoa, and Turin as all the previous ones). Curcio and Franceschini from prison immediately declared it a fake (CdS 19.4.1978). The true communiqué n.7 was issued by the BR a few days later and its authenticity certified by the standard template and two pictures of Moro holding a newspaper dated after the issue of the forged document (Biondo and Veneziao 2008; Moretti 2000, 139).

Conclusions

As it should be clear from this chapter, political agents who resort to violence are not just marginally concerned with communication, but they design violence itself taking communicative implications into consideration. Some do not even bother with violence at all, and parasitically advertise their group on the back of the violence carried out by others. In spite of its small scale and short duration, the Italian conflict that we examined is replete with examples of the varied dynamics that characterize the interplay between political violence and communication, first and foremost the management of group identity signals.

A plausible theoretical expectation in asymmetric conflicts is that the weaker side, by attacking the stronger side, aims to impact not on unattainable military objectives as in a conventional war, but rather pursue 'propaganda by deed'. This includes a menu of intermediate objectives: destabilize the enemy by scaring its supporters; show the enemy's weakness and induce it to embark on a costly repression bound to foster wider revolt; gain a reputation for self-abnegation and resilience in the service of a cause and thus lure new recruits. These aims are, one would expect, more efficiently pursued if the perpetrators are recognizable by some sign, notably a group name, and clearly stick to it across attacks.

This expectation, however, is far from universally borne out. At the beginning of the Italian conflict especially, we found considerable semantic turbulence: forty-five per cent of violent attacks recorded in the period we examined were never claimed. More baffling still, seventy-one per cent of claimed attacks (198) were claimed under a name used only once. An odd cacophony, the exact cause of which is hard to isolate: maybe some groups had the lifespan of fruit flies; or were busy experimenting with names to see what might catch on. Or the same perpetrators thought of names as flags, the more one waves, the larger and scarier one's camp appears to the enemy. When there is only a weakly organized political camp violently defying the state and its members are not too concerned to maintain a coherent strategy of attacks, violence is attributed to the work of that camp even if unclaimed or claimed under a plethora of names.

This approach, if indeed it was an approach, was, however, short-lived. On the one hand, perpetrators want their personal identity to remain veiled, and in this regard their enemies are the incumbents against whom they fight and who try to catch them. On the other hand they want the actions they carry out to be duly attributed to their group, and in this respect their enemies are different. They include competing groups from their same political camp who want to acquire credit for actions they did not carry out; and enemies from the opposite camp, sometimes in collusion with the incumbents, and criminals. Violent extremist struggles often fail to achieve their political objectives (Fortna 2015), but those not versed in communication fail faster.

As the conflict progressed and the seriousness of the attacks increased, various challenges led the main left-wing group, the Red Brigades to close ranks and align their behaviour with the expectation as they grasped the importance of claiming and enforcing their 'property rights' over their attacks, which meant to be able to enforce the property rights over their group identifiers. As the BR gained a dominant position their role in the dynamics of group identity signalling became a model, which attracted both types of mimics, the free-riders who did nothing but falsely sign with their group name action carried out by BR and the false-flaggers who carried out attacks falsely claiming them with the name of BR.

The information on the signalling warfare that we managed to unearth through data and biographies reveal that the BR responded to mimics, actual or potential, by designing claiming procedures and identifiers such that neither free-riders nor false-flaggers could afford to mimic. The BR claims were sheltered from mimics by carefully choosing the timing of the claim, the wording of the claim, the technology employed in the issuing of the claim, and the details of the attacks mentioned in the claims. All these items shared the key requirement stipulated by signalling theory that must be met to transmit credible information: they could only be known, possessed, or enacted by the real perpetrators. In addition, they resorted to offering an indirect 'signature' of their identity by using the same weapon.

This gamut of strategies was effective against free-riders, in both preventing false claims and counterclaiming ex-post. False-flaggers were harder to fight, and not just because it is easier to prove 'we did it' than 'we did not do it'. Those of the criminal type saw their opportunities to exploit the intimidation power of the BR when the BR resorted to kidnapping and robberies to fund their campaign—denials were less credible as mimicry could (and did) go both ways as the BR themselves tried to blame criminals for what they had actually carried out. False-flaggers of the political type were to some extent constrained by the choice of target: if they chose targets that were of a plausible left-wing kind they might end up doing the work for the BR, while if they chose right-wing targets, they risk harming their own side. Right-wing extremists went for the remaining option, namely to attack indiscriminately innocent people, but they could only blame these acts on loser left-wing groups such as the anarchists. They could not easily blame them on the BR who had publicly vowed to abstain from bombing precisely in order to thwart attempts of that nature.

References

Anonymous. 1979. 'Phenomenological and Dynamic Aspects of Terrorism in Italy'. *Terrorism* 2 (3–4): 159–170.

Bacharach, Michael and Diego Gambetta. 2001. 'Trust as Type Identification'. In *Trust and Deception in Virtual Societies*, edited by Cristiano Castelfranchi and Yao-Hua Tan, 1–26. Dordrecht: Springer.

Bale, Jeffrey M. 1987. 'Right-Wing Terrorists and the Extraparliamentary Left in Post-World War II Europe: Collusion or Manipulation?'. *Berkeley Journal of Sociology* 32: 193–236.

Bale, Jeffrey M. 1996. 'The May 1973 Terrorist Attack at Milan Police HQ: Anarchist 'propaganda of the Deed' or 'false-flag' Provocation?'. *Terrorism and Political Violence* 8 (1): 132–166.

Barbieri, Paolo and Paolo Cucchiarelli. 2003. *La strage con i capelli bianchi. La sentenza per piazza Fontana*. Rome: Editori Riuniti.

Bianconi, Giovanni. 2003. *Mi dichiaro prigioniero politico. Storia delle Brigate Rosse*. Turin: Einaudi.

Bianconi, Giovanni. 2005. *A mano armata. Vita violenta di Giusva Fioravanti*. 5th edn. Milan: Dalai Editore.

Biondo, Nicola and Massimo Veneziano. 2008. *Il falsario di stato*. Rome: Cooper Edizioni.

Boraso, Giuliano. 2006. *Mucchio selvaggio. Ascesa apoteosi caduta dell'organizzazione Prima Linea*. Rome: Castelvecchi.

Bueno de Mesquita, Ethan, and Eric S. Dickson. 2007. 'The Propaganda of the Deed: Terrorism, Counterterrorism, and Mobilization'. *American Journal of Political Science* 51 (2): 364–381.

Cazzullo, Aldo. 2006. *I ragazzi che volevano fare la rivoluzione. 1968–1978. Storia critica di Lotta Continua*. Milan: Sperling & Kupfer.

Della Porta, Donatella. 1995. *Social Movements, Political Violence and the State: A Comparative Analysis of Italy and Germany*. New York: Cambridge University Press.

Della Porta, Donatella and Maurizio Rossi. 1984. *Cifre Crudeli. Bilancio Dei Terrorismi Italiani*. Bologna: Istituto Cattaneo.

De Palo, Graziella and Aldo Giannulli (eds). 1989. *La Strage Di Stato*. Rome: Edizioni Associate.

Enders, Walter and Todd Sandler. 2006. *The Political Economy of Terrorism*. Cambridge: Cambridge University Press.

Engene, Jan Oskar. 2004. *Terrorism In Western Europe: Explaining The Trends Since 1950*. Cheltenham: Edward Elgar Pub.

Fasanella, Giovanni and Alberto Franceschini. 2004. *Che cosa sono le BR. Le radici, la nascita, la storia, il presente*. Milan: BUR.

Ferraresi, Franco. 1996. *Threats to Democracy*. Princeton, NJ: Princeton University Press.

Fortna, Virginia Page. 2015. 'Do Terrorists Win? Rebels' Use of Terrorism and Civil War Outcomes'. *International Organization* 69 (3): 519–556.

Franceschini, Alberto, Pier Vittorio Buffa, and Franco Giustolisi. 1988. *Mara, Renato e io*. Milan: Mondadori.

Galleni, Marco. 1981. *Rapporto sul terrorismo*. Milan: Rizzoli.

Gallinari, Prospero. 2009. *Un contadino nella metropoli*. Milan: Bompiani.

Gambetta, Diego. 1993. *The Sicilian Mafia: The Business of Private Protection*. Cambridge, MA: Harvard University Press.

Gambetta, Diego. 2005a. 'Conclusions'. In *Making Sense of Suicide Missions*, edited by Diego Gambetta. Oxford and New York: Oxford University Press.

Gambetta, Diego. 2005b. 'Deceptive Mimicry in Humans'. In *Perspectives on Imitation: From Neuroscience to Social Science*. Volume 2: *Imitation, Human Development and Culture*, edited by Susan L. Hurley and Nick Chater, 221–242. Cambridge, MA: MIT Press,.

Gambetta, Diego. 2009a. *Codes of the Underworld: How Criminals Communicate*. Princeton, NJ: Princeton University Press.

Gambetta, Diego. 2009b. 'Signalling'. In *The Oxford Handbook of Analytical Sociology*, edited by Peter. Hedström and Peter S. Bearman. Oxford: Oxford University Press.

Ganser, Daniele. 2005. *NATO's Secret Armies: Operation GLADIO and Terrorism in Western Europe*. London: Routledge.

Grandi, Aldo. 2005. *Insurrezione armata*. Milan: Rizzoli.

Grandi, Aldo. 2007. *L'ultimo brigatista*. Milan: BUR.

Harris, Robert. 2015. *Dictator*. London: Penguin.

Hoffman, Aaron M. 2010. 'Voice and Silence: Why Groups Take Credit for Acts of Terror'. *Journal of Peace Research* 47 (5): 615–626.

Hoffman, Aaron M., Crystal Shelton, and Erik Cleven. 2013. 'Press Freedom, Publicity, and the Cross-National Incidence of Transnational Terrorism'. *Political Research Quarterly* 66 (4): 896–909.

Kydd, Andrew H. and Barbara F. Walter. 2006. 'The Strategies of Terrorism'. *International Security* 31 (1): 49–80.

Lacqueur, Walter. 1997. *Terrorism*. London: Wiedenfeld and Nicolson.

LaFree, Gary. 2010. 'The Global Terrorism Database (GTD): Accomplishments and Challenges'. *Perspective on Terrorism* 4 (1): 24–46.

De Lutiis, Giuseppe. 1996. *Il lato scuro del potere: Associazioni Politiche E Strutture Militari Segrete Dal 1946 Ad Oggi*. Rome: Editori Riuniti.

Marighella, Carlos. 1969. 'Mini Manual of the Urban Guerrilla'. http://archive.org/details/Mini-manualOfTheUrbanGuerrilla (accessed 1 June 2015).

Marletti, Carlo et al. 2004. *Il Piemonte E Torino alla prova del terrorismo*. Soveria Mannelli: Rubettino.

Min, Eric. 2013. 'Taking Responsibility: When and Why Terrorists Claim Attacks'. Paper presented at the 2013 Annual Meeting of the American Political Science Association, Chicago, IL, 22 August.

Moran, Sue Ellen. 1986. *Court Depositions of Three Red Brigadists*. Santa Monica, CA: RAND Corporation.

Moretti, Mario. 2000. *Brigate Rosse: Una storia italiana* (3rd edn). Milan: Baldini & Castoldi.

Moss, David. 1989. *The Politics of Left-Wing Violence in Italy, 1969–85*. Basingstoke: Macmillan.

Panvini, Guido. 2009. *Ordine nero, guerriglia rossa. La violenza politica nell'Italia degli anni Sessanta e Settanta*. Turin: Einaudi.

Pavone, Claudio. 1991. *Una guerra civile. Saggio storico sulla moralità nella Resistenza*. Torino: Bollati Boringhieri.

Peci, Patrizio. 1983. *Io l'infame. La mia storia da terrorista pentito*. Milan: Mondadori.

Pisano, Vittorfranco S. 1979. 'A Survey of Terrorism of the Left in Italy: 1970–78'. *Terrorism* 2 (3–4): 171–212.

Podda, Stefania. 2007. *Nome di battaglia Mara. Vita e morte di Margherita Cagol il primo capo delle Br*. Milano: Sperling & Kupfer.

Progetto Memoria. 2007. *La mappa perduta*. Dogliani: Cooperativa Sensibili alle Foglie.

Rapoport, David C. 1997. 'To Claim or Not to Claim; That Is the Question—Always!'. *Terrorism and Political Violence* 9 (1): 11–17.

Raufer, Xavier. 1993. 'The Red Brigades: Farewell to Arms'. *Studies in Conflict & Terrorism* 16 (4): 315–325.

Rimanelli, Marco. 1989. 'Italian Terrorism and Society, 1940s–1980s: Roots, Ideologies, Evolution, and International Connections'. *Terrorism* 12 (4): 249–296.

Rohner, Dominic and Bruno S. Frey. 2007. 'Blood and Ink! The Common-Interest-Game between Terrorists and the Media'. *Public Choice* 133 (1–2): 129–145.

Ruggiero, Lorenzo (ed.). 2007. *Dossier Brigate Rosse 1969–1975*. Milan: Kaos.

Sánchez-Cuenca, Ignacio. 2007. 'The Dynamics Of Nationalist Terrorism: ETA and the IRA'. *Terrorism and Political Violence* 19 (3): 289–306.

Segio, Sergio. 2006. *Una vita in Prima Linea*. Milano: Rizzoli.

Spence, Michael A. 1973. 'Job Market Signaling'. *Quarterly Journal of Economics* 87 (3): 355–374.

Tessandori, Vincenzo. 2004. *BR. Imputazione: banda armata*. 4th edn. Milano: Dalai Editore.

Weinberg, Leonard. 1995. 'Italian Neo-fascist Terrorism: A Comparative Perspective'. *Terrorism and Political Violence* 7 (1): 221–238.

Wilkinson, Paul. 1997. 'The Media and Terrorism: A Reassessment'. *Terrorism and Political Violence* 9 (2): 51–64.

Wright, Austin L. 2011. 'Why Do Terrorists Claim Credit?'. Working Paper. https://www.princeton.edu/politics/about/file-repository/public/Wright_CreditTaking_11.3.2011.pdf (accessed 1 June 2015).

6

Where Are the Mimics When Passing Seems Easy?

The Rwandan Genocide in Comparative Perspective

David D. Laitin

There is a curious tension in reading Diego Gambetta's (2005) field-framing treatise on deceptive mimicry.[1] The reader is led to believe that the tactics of the mimic are ubiquitous and quotidian. In the short space of two months of news stories in New York, he reports, deceptive mimicry crawled into Christmas giving, parcel delivery, electrical repairs, and drug enforcement. No wonder, he concludes, 'the list of mimicry episodes becomes endless' (Gambetta 2005, 223).[2] Yet caveat emptor! Gambetta does not report on the denominator, for example, the total number of interpersonal interactions that took place in New York over that same two-month period. Surely, those in which someone was fooled through the use of deceptive mimicry constitute a miniscule percentage. He tells us, 'often mimics succeed', (p. 241) but again without any attention to the denominator or to compute the number of times potentially rewarding mimicry is not even attempted.

In the case of the Holocaust, we do have a sense of the denominator. Weitzman (1999) is referenced (p. 232) for an estimation of 'thousands of Jews'—mostly women—avoiding the gas chambers for successfully posing as *goyim* (in this context, members of the German nation).[3] Riggs (2002) is referenced (p. 238) for an estimation of some 1200 Jews who avoided concentration camps through a more outrageous stratagem, namely, volunteering

[1] This paper was originally prepared for a conference organized by Diego Gambetta on 'Mimicry in Civil Wars: The Strategic Use of Identity Signals' 7–8 December 2007, held at the Collège de France, Paris with no substantive updates since 2010. Thanks to Jeremy Weinstein for his useful comments on a pre-conference draft; to Michael Biggs, Jon Elster, James Fearon, and Andrea Patacconi for comments during the conference; to Jessica Gottlieb for research assistance; to Risa Kitagawa for uncovering a useful dataset; and to Scott Straus for his careful read of the Rwanda case study.
[2] Readers of Stephen Holmes (this volume) will be similarly impressed by the ubiquity of mimicry during the chaotic period of the Iraq War (2006 and years following), totally blurring the identity boundaries of the Shia/Sunni divide.
[3] Quoted in Weitzman (1999).

David D. Laitin, *Where Are the Mimics When Passing Seems Easy?* In: *Fight, Flight, Mimic.* Edited by: Diego Gambetta and Thomas Hegghammer, Oxford University Press. © David D. Laitin (2024). DOI: 10.1093/9780191802454.003.0006

in the Nazi army. But a few thousand in the numerator with an estimated six million in the denominator makes mimicry, sadly in this case, a rare event indeed. Of course, we have not counted those who tried and failed, but we can infer that European Jews did not believe, in their threatening environment, that 'often mimics succeed'.

Gambetta of course is aware that the dupe—the person the mimic is trying to fool—is not invariably duped (Gambetta 2005: see fn. 6, p. 229). From a theoretical perspective, his exposition of the equilibrium outcomes of semiotic warfare predicts that models (of the culture that is to be mimicked) will hone signals that are easy enough for group members to emit yet difficult enough for mimics to send such that a semi-sorting equilibrium would result, in which successful mimicry becomes the exception in social interactions. On a more macro level, Gambetta raises the possibility of boundary protecting collusion (whether intentional or not) between the leaders of the group from which the mimic comes and the group whose members the mimic seeks to impersonate that jointly works to limit successful mimicry.

And Gambetta's historical example of the Nazi Holocaust similarly reveals an example of failure. Polish Jews in the 1930s, although they lived in close proximity to Poles and were passively exposed to Catholic symbolism, generally failed to convince the administrators of the Holocaust that their signals of being Polish were genuine. Gambetta infers that the Poles had developed several cultural traits that were not easily copied by Jews, and Jews (especially Jewish males) developed cultural repertoires that made them look very un-Polish. These cultural signals, often inadvertently sent, made it relatively easy for Nazis (the dupes) to detect Jews-posing-as-Poles (the mimics). Under these conditions, it is unlikely there would be an abundance of mimicry attempts.

Fearon (in this volume) formalizes the conditions under which we would observe successful mimicry, and by so doing provides a model for the limits to its use. His model considers the value of passing for the mimic; the cost to the mimic if caught; the effort the mimic must expend for a reasonable chance of success in fooling the dupe; the potential dupe's cost for investigating as well as the dupe's loss for failure. In examining these parameters, Fearon shows that under some parameter values it is not worth the dupe's effort to investigate and therefore we should observe mimicry wherever it pays off for the mimic. Under another set of values, the dupe is sufficiently motivated to investigate everyone, and then we should observe mimicry only rarely, and only for those potential mimics whose costs of effort are sufficiently low. There is an intermediate set of values where the dupe investigates only a subset of possible cases of mimicry. One implication of this case, something to which I shall be returning, is that as the stakes go up for success (failure) to the mimic (dupe), we should observe lower rates of mimicry. The reason is that when the stakes are high, mimicry attempts must be rare or else the 'dupes' would investigate more often or more thoroughly, thereby increasing the degree of effort or skill by potential mimics necessary for success.

Indeed, as I shall show, when the stakes could not be higher (genocide) and the cultural differences between the mimic and dupe small, the observed cases of life-saving mimicry are rare. More specifically, the paper seeks to outline factors limiting the use of *defensive* mimicry (that is, mimicry to save oneself from capture, torture, or execution) under civil war conditions. But exposing these limits does not imply the irrelevance of mimicry under such conditions; rather it implies that the threat of mimicry creates incentives for the potential dupes to construct mimicry-resistant institutions, a key element in the larger process of state-building.

The plan of this chapter is as follows. In the first section, relying upon evidence from civil wars in the past half-century, I debunk the stylized fact—through an attention to the denominator—of a high prevalence of mimicry under conditions for which mimicry, if successful, would be life-saving. In the second section, relying on quantitative data from the Minorities at Risk dataset, I show that cultural proximity between a minority and a majority group does not reduce the probability that the minority would be embroiled in a civil war. These data imply that under conditions supporting a civil war onset, threatened minorities who through cultural proximity are more easily able to pass as members of the majority do not take advantage of this opportunity to avoid victimization as much as one would expect given the high costs of being exposed as mimics. The third section examines the case of genocide in Rwanda (1994) to illustrate the general points made in the earlier two sections. In the fourth section, a variety of explanations are offered. To answer the question of why cultural proximity is not an easy invitation to mimicry, I build upon an insight from game theory which suggests that cultural proximity should not predict the ease of successful passing or mimicry; rather it should predict the construction of institutions to prevent it. To answer the question of why we do not see mimicry more generally in civil wars, I report on inconclusive results of Gambetta's conjecture on the joint production of ethnic boundaries. I then emphasize the importance and relevance of Gambetta's distinction between offensive and defensive mimicry. In civil wars, escape from threat requires defensive mimicry, which presents to the mimic a truly heroic challenge, and thereby limits its use and effectiveness.

Debunking a Stylized Fact

Kalyvas's *The Logic of Violence in Civil War* (2006) is the best micro-study available of civilian survival tactics in the face of civil war violence. In it, we read of strategies of flight from danger zones by the weak (without cross-dressing, as with Gambetta's father to escape the Nazis)[4], collaboration with the enemy (while maintaining your identity, as with the Kikuyu Home Guards in the Mau Mau rebellion in Kenya), seeking security within newly defined cleavage lines (discussed in Kalyvas's section on 'Endogenous Polarization'), and denunciation of neighbours to demonstrate solidarity with the occupying forces. Elsewhere Kalyvas addresses a range of strategies adopted by civilians to hide their true preferences, such as double-dealing and hedging that become aspects of daily life when all society is 'riddled with deceit'.[5] But there is no mention of individual strategies of the weak to mimic the strong; rather the weak are trying to make the strong think they are something else (not the strong, though) that could improve their chances of not being targeted.

To be sure, without a theoretical expectation of a stratagem, even the best field researchers can be blind to its presence. But the absence of such examples in a book cataloguing the tactical range of civilian survival possibilities remains a troubling datum. Likewise, in his extensive review of deadly ethnic riots, Horowitz (2001) is impressed by the low degrees to which Type A errors (killing members of your own group, mistaking them for the other) and Type B errors (setting members of the other group free in your belief that they are in your group) are reported.[6] This paper will therefore focus on the other side of mimicry's coin—identifying its limits when the rewards for success are a matter of life and death.

Cultural Proximity and Civil War Violence

Any strategic theory of mimicry would foresee its probability of use under conditions of threat to be higher to the extent that the cultural repertoires of

---

[4] Diego Gambetta, pers. com.
[5] Kalyvas is quoting from M. Dillon (1990, 299).
[6] This report is astonishing given recent experimental work (Habyarimana et al. 2009, 48–52). In experiments conducted in Uganda, the authors found that although co-ethnics had a marked advantage in placing one another, their subjects nonetheless miscoded in-group members 33% of the time. If these experiments have external validity, one would think that riots and genocides would be fraught with errors—killings of one's ethnic kin through misidentification, and missing targets through mimicry.

mimic and dupe overlapped (Fearon's parameter 'e', this volume). This lowers the cost of dissimulation. We should therefore expect, if we examine cases where minorities are at risk of repression from dominant populations, where they could not conceive of winning in a fight, and where they lacked resources for flight, that they would assimilate the repertoires of the dominant group to avoid violent repression. As a first step in thinking about the empirical relationship of ease of boundary crossing and use of mimicry (through assimilation), this section will employ data from the Minorities at Risk (MAR) dataset in order to show that cultural proximity does not limit the chances of violent conflict in which minorities pay heavy costs.[7]

The MAR dataset is organized with an ethnic minority and the country in which this minority lives as the unit of observation. For example, 'Kurds/Iraq' is one observation, 'Kurds/Iran' another, and 'Shi'ites/Iraq' yet another. James Fearon and I have embedded in that dataset a scale of the linguistic distance between the ancestral language of the minority group and that of the dominant group in the country. We find that there is no relationship between linguistic distance and the propensity of a minority group to be at war with the state. This challenges an assumption in Gambetta (in reporting on the Jews with regard to the Poles) and in Gellner (1983) with his notion of entropy-resistant cultural traits, that the closer the groups are culturally, the easier it will be for minorities to 'pass' as members of the majority group, and therefore the less violent group relations should be. Indeed, it is unproductive and confusing to militias representing the dominant group within the state to order the killing of an 'other' when sorting is problematical.

In the version of MAR I use for this section, there are 398 observations in the dataset. Each group gets coded on a range of social, economic, cultural, and political factors (as possible explanatory factors for rebellion) and a periodic score on rebellion, an ordinal variable going from 0 (no rebellion) to 7 (protracted civil war). From this variable, I code for each group/country the maximum value of rebellion in any year from 1960 to 2000. More than half the groups never had a rebellion against the state (and their maximum rebellion score is '0'). At the other end of the distribution, fifty-two groups had a full-scale rebellion, with a maximum rebellion score reaching '7'.

---

[7] The MAR dataset as originally developed, due to issues of selection, faced formidable problems in making inferences about susceptibility of groups to rebellion. Addressing these problems with considerable success, a new team is reformulating MAR, but as of this writing, it is not yet fully operational. Here I use an older version of the MAR. Although not corrected for selection issues, those concerns are less relevant here. The selection issues concerned the over-estimation of the probabilities of rebellion for all minority groups. But there was no theoretical worry that groups that were culturally similar to the dominant group would more likely be included in the dataset if there was violence compared to groups culturally different from the dominant group.

The first variable that proxies for cultural difference (and thus the cost of successful mimicry) is based on an algorithm for the linguistic difference between the dominant group in the country and the particular minority. The source of the data is from Ethnologue[8], which places all language groups in the world on a tree-like structure. Languages are first differentiated by the broadest set of families, and each family breaks down into smaller and smaller branches. In a sense, the authors of Ethnologue give each language group in the world an address. James Fearon and I coded this variable for the 298 language-based groups. For each country, we code the language of the dominant group politically, and then compute the number of branches each minority group shares in common with the dominant group. If the minority and the dominant group break away linguistically at the broadest set of language families (i.e. breaking away at the first break-off point), the variable takes on a value of 1 (for which 161 groups are coded). If the minority speaks the same language as the dominant group, we assign it a value of twenty (representing forty-one cases).

The second variable proxying for cultural difference compares the religion of the dominant group (if there is one) and the religion of the minority (again if there is one), and assigns a 1 if the religions are the same and a 2 if they are different. In the MAR dataset, 221 minorities could be coded as having a common religion. Of those 221, 107 (or forty-eight per cent) shared a religion with the dominant group in their country and 114 (or fifty-two per cent) had a different religion.

One of the startling recognitions as we began summarizing the MAR data was that high levels of ethnic difference between a minority and the dominant group did not correlate with higher rebellion scores. With regard to language distance, we found that the closer the address between the minority language and the dominant language, the higher the probability of rebellion. In a bivariate test, the data reveal that each step up the scale of language proximity raises the expected maximum rebellion score by 0.2 and this is significant at the ninety-nine per cent confidence level.[9] With full controls, and put into a regression model, this positive relationship is not robust (Laitin 2000). But in no way could we say that language distance makes violent conflict within

---

[8] Available at: http://www.ethnologue.com/.

[9] For this measure, James Fearon and I, relying on language trees developed by linguists for Ethnologue (see http://www.ethnologue.com/), have constructed a variable 'langfam' which measures the structural distance between the dominant group in the country and any particular minority within that country. The scale goes from 1 (meaning that the two languages are from different language families) to 20 (the two languages are the same). The pairwise correlation of langfam and the maximum rebellion score (from 1960 to 2000) for each group in the Minorities at Risk dataset (on a scale ranging from 0 with no violent conflict to 7 with full-scale civil war) is positive and has a value of .21 ($p = 0.004$).

states more likely. Similarly with religious difference. Here the data show no significant relationship at all in the bivariate correlation. Again, ease of passing does not relate to lower predicted probabilities of large-scale violence between a dominant majority group and a vulnerable minority.

These results are startling in part because of theories propagated by Huntington (1993) that 'civilizational divides' were dangerous. But they are startling as well for a deeper reason. Let us assume that a minority group faces oppression or at least discrimination. Under such conditions, rational members of the minority might seek to pass as members of the dominant group in order to enjoy the fruits of domination, or at least to avoid discrimination or oppression. Cultural proximity would ease that task. In consequence, with greater opportunity through passing for those with ambition, we should see lower levels of rebellion the easier it is to pass. In Hirschman's (1970) terms, if exit (from one's minority group) is so cheap, why should oppressed minorities choose voice (here meaning rebellion against the oppressive group)?

What to make of these results? Consider the list of cases from the MAR dataset where dominant and 'at risk' groups share a common language and a common religion, and find themselves against one another in a civil war.[10] Here we observe the Isaaq and Hawiye (Somalia), the Hutus (Burundi and Rwanda), the Tutsis (Rwanda), and the Palestinians in Jordan. The internecine wars among Somali clans (all Sunni Muslims and all speakers of Somali), between Hutus and Tutsis (both with similar religious attachments and speakers of Kirundi), and between Palestinians and the Hashemite-led state (both Sunni Muslims and speakers of Arabic) are clear examples of violent conflict when passing might be expected. The most troubling of these cases is that of Rwanda, where there was in 1994 a genocide perpetrated by the dominant Hutus against the minority Tutsis. Facing extermination and very low cultural barriers separating the groups, shouldn't we see (given Gambetta's conjecture on the availability of cultural signals) long-term investments in routines and paraphernalia to succeed in mimicry? In the next section, I will address this question.

Rwanda 1994: Where Was the Mimicry?

Rwanda has had a bloody history. The minority Tutsis were the colonial favourites under Belgian rule, but a Hutu Revolution in 1959, on the eve of independence, upset the cultural division of power. That revolution, set

---

[10] For those who request the dataset, the command for this list is: list group country if langfam==20 & Religdif==1 & maxreb60>3 &maxreb60~=.

off by an attempted assassination of a leading Hutu nationalist figure, led to a massacre of at least a thousand Tutsis, and by some accounts, many more. Violence against Tutsis recurred in 1963–1964. From independence onwards, the dominant (by population) Hutus remained in power, but in the early 1990s faced a rebellion of the Rwandan Patriotic Front (RPF) led by the sons of the refugees of the 1959 massacres. In this context, the assassination of Rwandan President Habyarimana (a moderate in the context of leadership in the MRND, the most anti-Tutsi of the main Hutu-dominant parties) in April 1994, presumably (but without clear evidence) by Tutsi rebels, was the proximate cause of the mass killings of Tutsis and pro-peace Hutus—in which from a half to three-quarters of a million were massacred—mostly organized by the radical Hutu militias known as the *Interahamwe*.[11]

In the many books, articles, and testimonies of perpetrators and survivors of the genocide in Rwanda of 1994, a wide range of survival stratagems is revealed: bribery, cultivating Hutu connections, flight, seeking refuge (in churches, UN offices, and a five-star hotel), and hiding. Yet Tutsis seeking to pass as or mimic Hutus—for example, through the printing of counterfeit ethnic ID, the principal piece of evidence Hutu-led militias relied upon—was barely mentioned.[12]

This is puzzling for a host of reasons. First, Tutsis faced horrific consequences of being identified as a Tutsi in Rwanda in 1994. Second, because of the great cultural similarity between the groups, passing should have been relatively easy. Third, with extensive exogamy between the groups (especially in the cities), there were great zones of ambiguity as to who was a Hutu and who a Tutsi. This was especially true in the southernmost province of Butare, on the border with Burundi, yet an extensive report on the genocide in Butare provided no examples of deviousness when it came to ethnic identities. Since the genocide was carried out in no small part by northern militias, it is surprising that there is no evidence of southern Tutsis fooling northern militia heads of their identities (Guichaoua 2005 see esp. 265).[13] Fourth, many Tutsis looked closer to the Hutu physical prototype than they did to the Tutsi,

---

[11] The *Interahamwe* was the youth wing of the MRND that became the principal Hutu paramilitary organization, with full government backing, that implemented the genocide. See the Wikipedia page, http://en.wikipedia.org/wiki/Interahamwe, for full details.

[12] De Vulpian (2004, 130); Scott Straus (pers. comm. 7 October 2009) speculates that the stylized fact driving this section of the paper, namely the low level of attempted mimicry, may be due to a selection bias—Tutsis may have tried to pass, but were killed when they did, and we have therefore no reports of their attempts. This is possible but not entirely convincing, as many other failed strategies came to light after the genocide.

[13] Scott Straus (pers. comm, 7 October 2009) pointed out that while the term *interahamwe* technically refers to the youth wing of the MRND, many units that became armed militias in 1994 were mostly recruited from the north. Still, he added, nearly 90 per cent of the killings involved local militias versus local populations. So cross-regional confrontations between genociders and their victims were rare.

and vice versa. Identification was hardly foolproof.[14] Fifth, the danger signals were clear for many months that genocide was in the works, so there was time for the planning and implementation of short-term survival strategies. And sixth, Tutsis had (from the Hutu Revolution of 1959) excellent historic precedent to expect future massacres (Gourevitch 1998, 23). Tutsis therefore had some thirty-five years to prepare longer-term survival strategies—e.g. create a market for fake IDs—in the likely chance of a new set of massacres.[15]

Paul Rusesabagina (2006, 98), who became well-known as the manager of 'Hotel Rwanda', writes that anyone was in trouble who could not 'prove their Hutuness'. So if he is right, we have a case where three-quarters of a million people who shared a language, a religion, and a way of life with Hutus could not figure out how to furnish proof that would assure their survival! It is clear that Tutsis were quite resourceful in carrying out survival strategies. Indeed, there is a standard narrative among Tutsis on how they survived previous massacres. Then incumbent president of Rwanda, Paul Kagame reported to Gourevitch on his family's escape from Rwanda to Uganda in the wake of the Hutu Revolution. Some of them, Kagame reported, through bribery and forgery that furnished proof of birth in Uganda, got Ugandan citizenship papers (Gourevitch 1998, 211). Kagame's family proved its Ugandaness. But faking Hutuness in 1994 was rarely mentioned as we shall see in the catalogue of survival strategies noted below.

Flight

Flight was the principal strategy with good historical precedent. In 1960, after the 1959 massacres, there was a mass exodus of Tutsis to Uganda. Many of these refugees and their children became de facto Ugandans, English-speakers, and formed a backbone of Yoweri Museveni's revolutionary army that captured state power in Uganda. In the years preceding the genocide in Rwanda, many Tutsi families sent their children to relatives in Butare (the province in southern Rwanda which had been a zone of safety in the earlier massacres, but alas not the one of 1994) and across the southern border to Burundi (de Vulpian 2004, 129–136) where Tutsis controlled the state apparatus.

There are theoretical and practical reasons pointing to flight. Kalyvas (2006) has shown that local feuds map closely on to civil war dynamics,

---

[14] There were reports of Tutsis considering themselves lucky for looking like the Hutu phenotype—indeed the literature emphasizes that despite the ideology of phenotypic difference, this was not an accurate way to sort the population, even for insiders—and thereby avoiding murder during an attack.

[15] State ID cards were eminently forgeable, and required a mere checkmark before the label 'Hutu' or 'Tutsi' to establish ethnic identity. For examples of these cards, see http://www.preventgenocide.org/edu/pastgenocides/rwanda/indangamuntu.htm.

and once the genocidal project appeared dominant from the centre, local Hutus had a powerful incentive to ally with the dominant political coalition's identification of the enemy, or be denounced. Given local Hutu incentives, it would be rational for Tutsis to choose flight over an effort at deception, especially in densely populated and heterogeneous areas, and under conditions where former friends (who knew each other's ethnic roots) had incentives to turncoat.

Hiding

Hiding stories abound in the literature. One report is of a family living above a fake ceiling in the home of a Hutu friend (de Vulpian 2004, 129–136). Rusesabagina (2006, 96–97) reports of Tutsis 'running to places where they thought they might be spared. Churches were favourite hiding places … ' as they had been safe refuges in 1959. Abbé Modeste hid for weeks in his sacristy, under the desk in his study, and ultimately into the rafters at the home of neighbouring nuns (Gourevitch 1998, 21). Samuel Ndagijimana sought refuge in a Seventh-Day Adventist mission (Gourevitch 1998, 25). United Nations barracks, at least as long as the UN force saw its goal as protecting citizen lives, were also a place for Tutsis to congregate and hide.

The more usual move for many Tutsis, devoid of friendships that they could exploit in a time when all-too-many Hutu friends were quickly becoming turncoats, was to flee to the mountains, to the caves, and also to the marshes (Gourevitch 1998, 30).

Hiding was hardly a one-time proposition. François Xavier Nkurunziza, a Kigali lawyer with mixed parents, went from one hiding place to another (Gourevitch 1998, 23). One well-known Tutsi deputy hid in a chicken coop for three days, before trying to escape in the direction of Burundi, and remained for days in the marshes separating Rwanda from Burundi, before being identified and shot dead with her own pistol.

In an account of fourteen Rwandan survivors, Jean Hatzfeld (2000) reports on the way Tutsis escaped genocide in the town of Nyamata in the Bugesera district. All survivors recounted some method of hiding. One young student was hidden and fed by a local Tutsi woman who was subsequently killed by her Hutu husband when he discovered this. Two female farmers hid in the woods and marshes to survive. One of them reported always having been segregated from Hutus in their village yet with a *modus vivendi* between them that only broke down in the months preceding the genocide. One male shepherd hid in the marshes with the other Tutsis who survived the initial massacres in churches. One woman was hidden by a Hutu neighbour, but he kicked her out when local authorities failed to find her body among the corpses in the village, thereby arousing suspicion. She was then dropped off in a forest where she made her way to a refugee camp just over the Burundi border. According to another article citing escape narratives (Hazan 1998), one Tutsi hid in a church, one hid in her housekeeper's home in a sack of charcoal, and another fled in the woods on foot.

Bribery

Rusesabagina (2006, 90–91) moved his entire family from the hotel he managed to the hotel that would bring safety, and the key to success was the bribery of Hutu police. Gourevitch has extensive material on a Tutsi doctor, Odette whose husband had a Tutsi father and Hutu mother and was able to procure Hutu identity papers. He and Odette at the beginning of the extermination were paying $300 a day in protection to neighbourhood police, and were hiding Odette's sister, a deputy in the parliament (p. 72).

Cultivating Hutu Friendships

Rusesabagina (2006, 100) emphasizes the strategy of relying on pre-genocide networks to help with flight or with hiding. His most valuable possession at the start of the killings was his extensive address book of Hutus to whom he had given help over the years, and begging them for a range of protection services. Gourevitch (1998, 22) reports on many friendships that helped Tutsis survive. Lurent Nkongoli told him that he was saved by a Hutu who told him to 'run away, we don't want to see your corpse'. Félix Semwaga, a Hutu local authority in the commune of Mbazi in Butare Province, sought to stay the killings but the new bureaucratic appointments coming from the political center to Butare overruled him. At best, the communal authorities in Mbazi held out longer than most other communities. Semwaga himself hid many Tutsi neighbours in his home. But eventually his house was attacked and the refugees were killed (Guichaoua 2005, 288–290).

Narratives of Passing and Mimicry

What about stories of mistaken identities and passing? It should be emphasized that these moves were in the feasible strategy set of potential victims. Indeed, there was passing among Tutsis well before the genocide. Following the Hutu Revolution in 1959, the percentage of Tutsi in the Rwandan population declined sharply, partly because many had been massacred or fled, and partly because some found ways to redefine themselves as Hutu (des Forges 1999, 40). Des Forges provides other examples, most of them requiring the support of (Hutu) local authorities. During the genocide, she reports, persons who hoped to pass for Hutu often 'lost' their identity cards and requested temporary papers from the councillor or a new card from the *bourgmestre* (the mayor, or communal leader), hoping the administrator would be persuaded to falsify the document. An International Criminal Tribunal of Rwanda (ICTR) testimony by one bourgmestre stated, 'In the countryside, the mere fact of giving an attestation to a person sufficed to save him' (des Forges 1999, 238). Similarly, there are scattered reports of fake identity cards issued in Butare obtained with the help of those Hutu mayors who took the risk of assisting Tutsi friends or relatives (des Forges 1999, 693). There is evidence as well that those with mixed heritage may have been more likely to try to pass. Hutu mothers of children fathered by Tutsi sometimes tried to protect their children by claiming they were illegitimate and seeking to have them registered on their cards as Hutu rather than basing identity on the cards of the husbands of their mothers (des Forges 1999, 238).

One pass, from a Human Rights Watch report (1995), was spontaneous, done within moments of an imminent attack: 'A young Hutu woman lent her identity card to a Tutsi so that she could pass barriers on her attempted flight from death'. Gourevitch (1998, 72) provides another, one that required planning. The deputy who was shot in the marshes made her initial escape (after days in a chicken coop) in a car where she wore a headscarf impersonating a Muslim, and this achieved success, since the driver of the car (Odette's husband) had a Hutu-identity card and claimed to be escaping from the RPF. Odette's survival was due in part to mistaken identity. The killers had tracked her down to the hospital at which she worked, but they found there a Hutu colleague of hers, also named Odette, who was taller than she (i.e. looked like a Tutsi), and they tortured and killed her instead (Gourevitch 1998, 82–83). These stories are suggestive, but how common were they?

Exploiting a Dataset of Survival Strategies

From the Outreach Programme on the Rwanda Genocide we have seventy-eight testimonies that provide evidence of survival strategies of vulnerable Tutsis and can provide a quantitative assessment to the question of how common was mimicry in the Tutsi genocide. Nearly all of the testimonies are from women survivors who responded to the offer of outreach and in the course of receiving support, recorded their experiences in the face of mass killing. Most of the narratives provided to the organization reported on the efforts of the survivor's wider family to escape murder. I coded each narrative as to whether one of the strategies discussed above were employed.[16]

The set of strategies for survival included flight, fighting (often only with stones, leading to the question of why threatened Tutsis did not store guns), hiding (in lakes, in swamps, in sorghum fields), bribery (to get through a roadblock, or to be housed in a Hutu home), becoming a sex slave (that was usually accompanied by being given refuge in a Hutu house), getting refuge as a friend from a Hutu family, and getting refuge from other institutions (a church, the French, or Tutsi militias (the *inkotanyi*)). All these strategies were described as ad hoc reactions to the rumours and broadcasts oriented towards the Interahamwe volunteers (which, of course, the Tutsis could understand as well as the Hutus), suggesting that there was no pre-planned survival strategy for an often well-to-do social class that was under ultimate threat.

The strategies were impressively ad hoc. Among the testimonies, thirty-three (or forty-two per cent) reported that they took flight, seeking the border to Zaire, or to some place that might provide sanctuary, or to the forests and swamps where they could wait for the campaign to end. Twenty-nine (or thirty-seven per cent) report that they hid either outdoors (e.g. in a swamp), in a pile of dead bodies, or in their own compounds (often the latrine). Six testified that they bribed Interahamwe to let them through a roadblock, or to take them into their homes. Thirty-nine (fifty per cent) reported getting refuge in a Hutu home, though only in eight of these cases was this done out of compassion, while thirty-one of them involved becoming the 'wife' of the Hutu protector, i.e. a sex slave.

What about 'fight'? Interestingly, there were only seven cases where the men in a Tutsi compound tried to stand up to the Interahamwe attackers. In all reported cases, the Tutsis relied on stones and faced Hutus with machetes and guns. None of the men were saved through fighting.

---

[16] The data were collected in collaboration with an information and educational outreach program run by the United Nations Department of Public Information. See http://www.un.org/en/preventgenocide/rwanda/education/survivortestimonies.shtml. Concerning representativeness of this sample, it is obvious that looking at families with at least one survivor may add in a degree of bias. Certainly, since nearly all the testimonies are from women, the prominence of surviving through sex slavery is higher than for those females who were immediately hacked to death. But for the immediate reactions to the evidence of mass killing, there is no reason to think that these seventy-eight respondents were exceptional in any way. Since the descriptive statistics do not violate conventional reports on the strategy set, these data make possible informed guesses on the magnitude of each type of strategic effort.

Mimicry (or any form of dissimulation) was also rare, and mentioned in only five of the seventy-eight testimonies. None involved a pre-planned strategy of multiple identity cards. Alexandria reported that in her district, young women and girls were not killed, but incarcerated in a stadium and raped. When men were brought there for eventual murder, she testified, 'we tried to conceal the men by dressing them as women', but they were rapidly defrocked and murdered. Another woman, Bernadette, reported that her mother 'told us never to tell anyone that we were Tutsi, but to tell people we were Hutu'. Taking that advice, she went to a place where they were not known and told people their father was an officer in the army. However, the Interahamwe came to her house and seized her entire family. Catherine reports that at a roadblock she tore up her identity card, but that did no good as she was raped and turned into a sex slave. Eugenia also tore up her identity card claiming that she and the baby on her back were Hutus. She reports that no one believed her, but nonetheless she was cleared through a roadblock, and hid in a sorghum field until found by RPF soldiers. Mercia was given false papers by a Hutu soldier to serve as a nurse for his child, and meanwhile used her as a sex slave. Thelma lied about her identity at a roadblock and was let through. She got to a church ground, but was not permitted entry due to overcrowding, according to the priest. Hutu soldiers patrolled there and she was regularly raped while living on the church grounds. This dataset reveals rare cases of dissimulation, but hardly a mass effort that progressively blurred the boundaries separating Hutu from Tutsi.[17]

Towards an Explanation for the Absence of Mimicry in Rwanda's Genocide

Evidence exists that after the genocide, when Hutu perpetrators were being hunted down by the RPF and the ICTR, *Hutus* quickly learned the importance of passing. Guichaoua (Guichaoua 2005 264 fn. 13) reports, for example, on a Hutu murderer now living in France with a false identity card! Why not the Tutsis in the face of genocide?

There are two related questions that require explanations. First, why was the cultural proximity between Hutus and Tutsis not an invitation to greater levels of successful mimicry? Second, why is mimicry by the subordinate

---

[17] A sixth mention, with the genociders rather than the victims hiding their identities, concerned a group of young *interahamwe* who wore masks trying to hide their identities from Mercia's family whose men they were killing, as they were local kids well known to Mercia.

groups under civil war conditions less used than would be expected given the huge pay-offs for success?

Local Knowledge

As to the question about the cultural conditions that favour mimicry, a first cut answer is that cultural similarity is insufficient to reduce the barriers to passing when there is high local knowledge of where people fit into the social structure. Indeed, sharing a language and close social ties had an up and a down side for the safety of Tutsis. Gourevitch (1998, 65–66) relates a story of a Tutsi being made an honorary Hutu by villagers, but was later called in for having a fraudulent identity card and killed. This shows that mimicry into the group of the perpetrators (as opposed to Hutus convincing French authorities that they are Tutsis) is hard even when language, religion, and culture are the same. Local knowledge exposes this stratagem. In ethnically complex areas, there are (in the real world, as opposed to the world of field experiments) redundant methods of group identification, even when religion and language are the same. Tutsis seeking to portray themselves as Hutus in the *interahamwe* era seemed to face the same hurdles as Warsaw Jews seeking to portray themselves as Poles in the Nazi era, even if the cultural distances between perpetrators and victims were of different magnitudes.

Furthermore, the redundancy of ethnic markers makes passing dangerous (at least if it involves more than getting through a roadblock). In Somalia, where language and religion are the same across clans, the ability to produce credible genealogical sequences is a key sign of one's clan that is hard to fake, due in part to the potential follow-up questions on the clan's relationship to its eponymous founder, where all genealogies must reach, to make a credible claim to clan membership. (Getting that genealogical information drummed into you at an early age is crucial; much harder to learn this fluently for purposes of escape).[18] Cultural proximity, given redundancy of signs, therefore provides no open road to passing under civil war conditions.

Anti-mimicry Institutions

A second explanation for the failure of cultural proximity to foster physical security provides a game theoretic perspective to this case, and is consistent

---

[18] Having written many support letters for asylum status on behalf of Somalis, I have noted that the generation raised after 1991 no longer knows its genealogical sequence. This suggests that while clans remain important in Somalia, the clan structure as a system of common knowledge rules of social identity has collapsed. The implications of this for mimicry remain to be drawn.

with the MAR data presented in the third section of this paper.[19] If two groups are highly similar with easy passing, we should observe in equilibrium more institutions to raise the costs of mimicry. Thus mimicry should not vary across cases but what should vary are the social and institutional practices that regulate passing. In other words, boundary maintenance should vary. Indeed, what the Rwanda case demonstrates is the extensive work by the Hutu-led governments after the Hutu Revolution to create institutions (such as the continued reliance of colonial-era identity cards and the tight local control over personal registration) that would protect the boundary separating them from potential Tutsi passers. Cultural similarity therefore should not predict ease of passage, but rather a stronger institutional response to prevent it.

Considerable evidence supports this perspective. The literature notes the high levels of local control over identity in Rwanda. An average commune had between 40,000 and 50,000 residents and there were subdivisions within communes (sectors and cells) governed by party members that administered groups of 1000 people each. These same mayors and councillors were the primary implementers of the genocidal work decades later. But even where outsiders came into a community (say from the centre or from regional military posts or from neighbouring communities) these outsiders relied on local knowledge to find and identify victims. Locals knew who was who, from where they came, and who their parents were.[20] And when locals had doubts about the ethnic provenance of a family, they sought out elders or other knowledge-holders and deliberated about the ethnic origins of the family (Straus 2006). Also, prefects normally inspected commune-reported information carefully and could identify when there were fewer Tutsis reported than anticipated, indicating suspicion of passing or hiding (des Forges 1999, 239). An OAU report confirms that local officials were the organizers of most of the violence, so their ability to detect mimicry could have made the difference in strategy choice for Tutsis (Organization of African Unity 2000, 121).

Dating back to before the Habyarimana government, state authorities used the infamous identity cards to control and mobilize the population at the most local level. The control was implemented not just by the high ratio of officials to ordinary people but also by regulations governing population registration and movement. The Habyarimana government continued the

---

[19] Thanks to James Fearon and Andrea Patacconi who suggested this idea at the Collège de France conference and was subsequently formalized by Fearon for this volume.
[20] This is partially consistent with insights from Kalyvas (2006, 174–176) who emphasizes the importance of information provided to militias by locals. However, Kalyvas's prediction (pp. 218–220) that in zones of full control by a militia (in this case the Interahamwe) there would be low levels of killing (with denunciations by locals of who is an enemy being ignored by the militias) is not consistent with events in Rwanda.

use of identity cards and also required people moving from one location to another to register with the local authorities. Each commune submitted monthly, quarterly, and yearly reports of births, deaths, and movement into and out of the commune (des Forges 1999, 238). At the level of the *bourgmestres* (local government officials), there was Hutu solidarity: of the 143 *bourgmestres*, not one was Tutsi (Chrétien 1997, 84). Therefore, seeking sympathetic *bourgmestres* to get new identity cards would be a fool's journey for most Tutsi.

Still, the Hutu-led governments recognized the possibility of losing control over ethnic identification. Systematic threats were issued against falsifying one's identity well before the genocide. In 1973, there was a wave of anti-Tutsi persecutions that targeted Tutsi students and civil servants, mixed Tutsis and *troquers d'ethnies*, i.e. those who were determined to have fake papers (Chrétien 1995, 90). And since 1990, there were government authorized threats against 'les complices', or those Hutus who were complicit with the Tutsis who were suspected of hiding behind false papers. Warnings were issued to the Hutus that the Tutsis were trying to do this. One editorial claimed that eighty-five per cent of Tutsis had changed their ethnicity and that before befriending someone, readers were advised to ask their name and address so that their interlocutors could be sure they were a true Hutu. The Tutsis were incredibly promised 'no lying and no malice' (Chrétien 1995, 102–104). On the brink of the holocaust, a *Radio television libre des mille collines* (RTLM, the radio station that broadcast genocidal instructions to Hutu militias) broadcast on 28 May 1994 threatened those who authorities feared were using fake papers: 'If someone is in possession of false documents, if it is a member of the RPF, or if it is a well-known accomplice, there is no redemption' (Chrétien 1995, 193 author's trans.). There would be no redemption for those Tutsis nor their accomplices who tried to survive through documentary trickery. A cogent explanation for the immense attention by authorities in support of the integrity of the identification cards is that cultural proximity between Hutus and Tutsis added incentives to authorities to police the ethnic boundaries with greater vigilance.

But why *in general* is mimicry less evident among dominated minorities under conditions of civil war violence than Gambetta's initial statements would have led readers to believe? For Rwanda, the standard answer is state-centric. Hutu militias were provided with lists of targets that were prepared by local administrators for the Interahamwe. Targets were then lured out of their ad hoc hiding places with vague promises that standard sanctuaries, such as churches, would not be violated. Those who took the bait were assumed to be Tutsis and murdered en masse in those sanctuaries. On everyday patrol, militias would rely on physical markers and demand identity cards from those with Tutsi features (des Forges 1999; Prunier 1995, 244). A problem with this line of analysis is that it assumes that in the fog of war, ill-trained militias can see clearly. Are there not opportunities for targeted minorities to exploit the ignorance and the organizational failures of a congeries of militias set up by a weak government, generally the case in civil wars (Fearon and Laitin 2003), but even worse in Rwanda, which after the death of its president and ruled by an interim government, was in a state in turmoil?

Joint Boundary Policing

One intriguing answer to the question of boundary maintenance is the joint policing of mimicry by both sides in violent conflicts. The suggestion here is that in the Rwanda genocide of 1994, a joint production of self-defence (RPF and Interahamwe) gave incentives not only for Hutus to detect passing but for Tutsis to punish other Tutsis who sought to play the dangerous ethnic game that the Hutus were playing (Gourevitch 1998, 223). I have no data about the joint production of ethnic boundaries in civil war conditions, but in research on a related problem—why despised minorities don't pass as members of the 'cleaner' castes—among Scheduled Castes in India, Jews in early modern Europe, and Romas in Europe today—I recorded (in Laitin 1995) this mechanism of control as a normal part of everyday life.

The interests of the dominant group in society with regard to the erection of barriers to passing are obvious. If the sharing of the rewards of dominance are $1/n$ ('n' being the number of people who share the identity), then the rewards are smaller the more the group grows in size. Cultural barriers to detect and retard passing are therefore common in the social world. Brahmins going all out to police against marriage of their daughters with untouchables posing as caste Hindus; American WASPs setting the Humanities as a barrier to college entry to immigrants (noticed by Thorsten Veblen (1899) in his classic *Theory of the Leisure Class*); Hutus relying on identity cards to raise the barriers for Tutsis to pass. These examples of entry barriers are well known.[21]

The interests among members of the oppressed group, on the contrary, are not so obvious. But consider the situation of the 'sweepers' in India, who are consigned to the tasks of de-fouling the streets of villages and towns. Their

---

[21] See Barth (1969) for the classic statement of this cultural stratagem by the dominant cultural groups.

social status is at the very bottom of the ladder. But given that they are consigned to this role (and anyone seeking to perform it would lose all status), and given that the consequences of their withholding of their labour for the health of their towns are large, they have been able to secure attractive contracts with local authorities. Their economic returns are far higher than they would receive if they entered the bottom rungs of the non-despised occupational ladder. Therefore, they have an interest in protecting their economic niche by raising the cultural barriers between them and caste Hindus. Jews in early modern Europe (with banking services) and Romas in today's Europe (in performing seasonal labour) have had similar interests.

The joint production of barriers is easily observable in the political realm. Oppressed groups generate leaders who speak for their interests to the wider society. These leaders have an interest in being the monopoly representatives of their group. To the extent that rapid assimilation takes place, these politicians will lose their constituencies and the rents they receive from their monopoly brokerage role. Meanwhile, dominant groups seeking to protect their societies from uprisings by restive minorities will seek to identify, recognize, and negotiate with these monopoly mediators. In Germany and France today, for example, state authorities seek to co-opt and reward leaders of the 'Muslim' communities, and there are many Muslims who seek this role.[22]

The story of joint boundary production, while plausible as a general social phenomenon, may not provide much leverage for the Rwanda genocide. There is no evidence of *inkotanyi* reprisals against Tutsis who were complicit with Hutu authorities as a survival strategy. Moreover, post-genocide, the RPF government has been resolute to prevent any investigation of Tutsi behaviour during the genocide (Peskin 2008).

A more compelling answer to the question of why mimicry is not relied upon more under civil war conditions (as suggested at the Collège de France conference by Jon Elster) requires us to pay greater attention to Gambetta's distinction between offensive and defensive mimicry. In the former, you only need to succeed once for a successful heist. An excellent example are criminals in India posing as Naxalites to extort more efficiently, and this inducing the counter-move of 'brand protection' by the Naxalites (see Mamidi, this volume). In the latter, you need to succeed many times, as with the Tutsis trying to escape through a variety of road blocks. Even in the case of Iraq under conditions of state breakdown, where the prophet's remark in the *Hadith* 'Verily,

---

[22] The term 'monopoly mediators' is from de Swaan (1988); on the attempts to co-opt European Muslims, see Pfaff and Gill (2006).

war is deception' appears as an understatement, defensive mimicry was partially contained (Holmes, this volume). The more times a mimic needs to convince, the less the expected returns to mimicry.

Defensive mimicry is a challenge. At the beginning of the genocide, not all Tutsi were targeted, but being identified as someone trying to pass could easily make you a target. In Butare, it is reported that some Rwandans killed at the beginning of the genocide were officially Hutu on their identity cards, but someone had done research and had learned that they had 'previously' been Tutsi. Someone had gone to the home communes of those who were suspected to check on whether they were really Hutu or Tutsi (des Forges 1999, 474). Clearly, an initial success in passing did not assure security.

Consider the following report of a meeting in the commune of Nyanza in Butare province, one which included all the leading political and police figures, in which the topic was the threat of the RPF. The *bourgmestre* had a Hutu identity card but it was commonly known that his father was a Tutsi. For his moderate views, he was accused by the police chief of being a traitor, and fearing for his life, he took to hiding in the house of a communal policeman. The officers of the police force then went looking for him, failed to find him, went to his house and got no information there, so they murdered eleven members of his family. The *bourgmestre*'s hiding place was soon discovered by the genociders. He suspected their arrival and escaped. However, he was tracked down and arrested on the road. The militia tied him to their army vehicle and dragged him to near death and then killed him (Guichaoua 2005, 14, 262). Hutu identity cards, this sad vignette demonstrates, were not a failsafe mechanism.

Moreover, ID cards were not a sure-fire escape route, even for Hutus. In the commune of Nyakizu, Hutus at roadblocks armed were often arrested, detained, or killed, generally so they could be robbed by the powerful groups in the commune (des Forges 1999, 3). There was even a risk of ID cards being used for killing rather than for surviving. 'As the numbers of Tutsi were reduced', des Forges reports, 'the assailants deputized to kill them directed their violence increasingly against other Hutu ... Sometimes they confiscated identity cards from victims so that they could claim that [these Hutus] were Tutsi' (des Forges 1999, 572). For non-Hutus, a fake ID card could make you a target. The general point follows: if the ID card were merely used to gain quick entry for a rebel into a state property—this would be offensive mimicry—the chances would be good. But trying to use that same card to negotiate one's way through a myriad of road blocks and a plethora of police—and this is defensive mimicry—is no easy task.

Conclusion

If mimicry is a ubiquitous tactic in everyday life for criminals and parvenus, we should observe it in abundance in pogroms and civil wars where success would mean survival. And a signalling perspective would predict that we should especially see it under conditions where the combatants in communal violence share many cultural and physical features, lowering the barriers to successful passing. It is therefore surprising that in the Minorities at Risk dataset, assuming that minorities who are oppressed will try to pass as members of the dominant group thereby lowering the incentive for joining militias to fight against the state, we find no relationship between degree of cultural distance and likelihood of large-scale violence. The dataset reveals several telling cases where dominant and minority groups shared a language and religion and yet still fought bloody civil wars against one another. Most troubling is the case of Rwanda, in which as many as three-quarters of a million people identified by militias to be Tutsis were murdered. Standard accounts of Rwandan society underline that not only were cultural differences (language and religion) non-existent, but the often-cited physical differences were nowhere as stark as held by colonial administrators who had an interest—due to their divide to rule strategy—in reifying group differences. If cultural and physical barriers were low, this chapter asks, why do post-genocide testimonies so rarely mention mimicry as a common stratagem of survival? The absence of such accounts and the data from the MAR dataset suggest that mimicry might not be an available tactic for persecuted minorities under conditions of civil war. The purpose of this chapter was to raise this possibility, and to speculate as to why this dog did not bark.

It might first be objected that the testimonies available to me reflect a sample bias. Perhaps we are not hearing about the cases of successful deception because successful mimics remained silent, afraid of reprisals from both sides? I doubt that this is the case, because after the victory of the RPF there was no incentive for Tutsis to start a new round of reprisals, and every incentive to protect Tutsis from retribution, so the costs of admitting attempts at mimicry would be minimal.[23] Furthermore, journalists such as Gourevitch found in post-1994 Rwanda a desire by potential victims to give testimony.

---

[23] However, in the narrative dataset, young women who had become sex slaves reported that they faced rejection from their families and friends within the Tutsi community if they had borne a child of their tormentor (e.g. Adeline's testimony). One sex slave (Albertine) testified that 'even after the genocide, I stayed with him against my wishes, because I was ashamed of what everybody who knew me would say'. I don't know if successful mimicry evoked the same worry about social rejection post-genocide within restored Tutsi society. The few narratives that revealed mimicry were not apologetic, suggesting the social costs for in-group rejection due to mimicry were perceived by survivors to be low.

Unless there is a literature I am missing, I believe the general absence of (attempts at) mimicry under these genocidal conditions is a social fact that needs to be explained.

As suggested earlier, this social fact puts to some challenge notions proposed by Gambetta, but also to some recent field experimental work on ethnic identifiability, research that has shown mimicry to be a potentially valuable way to procure scarce goods (Habyarimana et al. 2009). Perhaps the experimental protocol revealing difficulties in detecting those posing as ethnic kin did not capture the essence of mimicry under civil war conditions? Unlike field experiments focusing on individual payoffs for correct guesses, what is striking about the vignettes provided here from the Rwanda genocide is how public the effort at deception had to be. The story had to be convincing not only to the militiamen, but to people in the locality who had an opportunity (and for Kalyvas 2006, the incentive) to expose the act of passing. Indeed road blocks in Rwanda were staffed by several monitors, each with different information, making it more difficult for the Tutsis to frame a story about themselves that would be cogent to the varied audience. The chance that one of the monitors might have information contradicting the story of the mimicker is higher the greater the number and diversity of monitors. In the Rwanda case, unlike the Habyarimana et al. (2009) experiments where dupes rarely updated with new information about prospective mimics, the Interahamwe murderers were highly motivated to fulfil a genocidal mission and surely paid greater attention to the quality of the message transmitted by those seeking to pass. Indeed, this chapter suggests that the external validity of the experiments did not travel well to civil war or genocidal contexts.

The material in this chapter therefore casts doubt on the feasibility of mimicry in civil war even when cultural conditions appear favourable. It supports two conjectures as to why this is the case. First, the defensive mimicry necessary for survival amidst genocide requires multiple successes across a range of auditors and over the course of time, and should be far less prevalent under civil war conditions than offensive mimicry. This seems an important distinction for future work on this topic. Second, the ease of passing should not predict the success of mimicry but rather the construction of institutions to prevent it. Here we see how Hutu authorities foresaw the threat of mimicry and built institutions to raise its costs.

In sum, an overview of a large-n dataset and of testimonies from the Rwanda genocide put to some challenge both theoretical and experimental work on the role of mimicry in social interactions, at least as applied to civil war conditions. Under these conditions, the pay-off to successful mimicry is extremely high, but the requirements for success (in passing through multiple tests) are also high, and the incentives for preventive detection (facilitated through formal and informal institutions) are perhaps even higher. The contribution of this study is its attention to the limits of successful mimicry even under conditions when the cultural costs of passing seem low.

References

Barth, Fredrik (ed.). 1969. *Ethnic Groups and Boundaries: The Social Organization of Culture Difference*. Oslo: Universitetsforlaget.

Chrétien, Jean-Pierre. 1995. *Rwanda, Les Medias Du Genocide*. Paris: Karthala.

Chrétien, Jean-Pierre. 1997. *Le Défi de l'Ethnisme*. Paris: Karthala.

Des Forges, Alison. 1999. *Leave None to Tell the Story*. New York: Human Rights Watch.

De Swaan, Abram. 1988. *In Care of the State: Health Care, Education and Welfare in Europe and the USA in the Modern Era*. New York: Oxford University Press.

De Vulpian, Laure. 2004. *Rwanda: Un Génocide Oublié? Un Procès Pour Mémoire*. Paris: Editions Complex.

Dillon, Martin. 1990. *The Dirty War*. New York: Routledge.

Fearon, James D. and David D. Laitin. 2003. 'Ethnicity, Insurgency, and Civil War'. *American Political Science Review* 97 (01): 75–90.

Gambetta, Diego. 2005. 'Deceptive Mimicry in Humans'. In *Perspectives on Imitation: From Neuroscience to Social Science*. Volume 2: *Imitation, Human Development and Culture*, edited by Susan L. Hurley and Nick Chater, 221–242. Cambridge, MA: MIT Press.

Gellner, Ernest. 1983. *Nations and Nationalism*. Ithaca, NY: Cornell University Press.

Gourevitch, Philip. 1998. *We Wish to Inform You That Tomorrow We Will Be Killed With Our Families: Stories from Rwanda*. New York: Farrar, Straus and Giroux.

Guichaoua, André. 2005. *Rwanda 1994: Les Politiques Du Génocide À Butare*. Paris: Karthala.

Habyarimana, James, Macartan Humphreys, Daniel N. Posner, and Jeremy M. Weinstein. 2009. *Coethnicity: Diversity and the Dilemmas of Collective Action*. New York: Russell Sage Foundation.

Hatzfeld, Jean. 2000. *Dans Le Nu de La Vie*. Paris: Seuil.

Hazan, Pierre. 1998. 'Des Rescapés Du Génocide Rwandais Témoignent'. *Le Temps*, 9 December.

Hirschman, Albert O. 1970. *Exit, Voice, and Loyalty: Responses to Decline in Firms, Organizations, and States*. Cambridge, MA: Harvard University Press.

Horowitz, Donald. 2001. *The Deadly Ethnic Riot*. Berkeley, CA: University of California Press.

Human Rights Watch/FIDH. 1995. 'Interviews from Butare, November 9, 1995'. Human Rights Watch/FIDH. http://www.hrw.org/legacy/reports/1999/rwanda/Geno11-4-04.htm#P469_153922.

Huntington, Samuel. 1993. 'The Clash of Civilizations?'. *Foreign Affairs* 72 (3): 22–49.

Kalyvas, Stathis N. 2006. *The Logic of Violence in Civil War*. Cambridge: Cambridge University Press.

Laitin, David D. 1995. 'Marginality A Microperspective'. *Rationality and Society* 7 (1): 31–57.

Laitin, David D. 2000. 'Language Conflict and Violence'. In *International Conflict Resolution After the Cold War*, edited by Paul C. Stern and Daniel Druckman, 531–568. Washington, DC: National Academy Press.

Organization of African Unity. 2000. 'Rapport Sur Le Génocide Au Rwanda'. Organization of African Unity. http://cec.rwanda.free.fr/documents/doc/Rapport_OUA/Rwanda-f/FR-14-CH.htm.

Peskin, Victor. 2008. *International Justice in Rwanda and the Balkans: Virtual Trials and the Struggle for State Cooperation*. New York: Cambridge University Press.

Pfaff, Steven and Anthony J. Gill. 2006. 'Will a Million Muslims March? Muslim Interest Organizations and Political Integration in Europe'. *Comparative Political Studies* 39 (7): 803–828.

Prunier, Gérard. 1995. *The Rwanda Crisis: History of a Genocide*. New York: Columbia University Press.

Riggs, Bryan Mark. 2002. *Hitler's Jewish Soldiers: The Untold Story of Nazi Racial Laws and Men of Jewish Descent in the German Military*. Lawrence, KS: University of Kansas Press.

Rusesabagina, Paul. 2006. *An Ordinary Man: An Autobiography*. New York: Viking.

Straus, Scott. 2006. *The Order of Genocide: Race, Power, and War in Rwanda*. Ithaca, NY: Cornell University Press.

Veblen, Thorstein. 1899. *The Theory of the Leisure Class: An Economic Study in the Evolution of Institutions*. London: Macmillan.

Weitzman, Lenore J. 1999. 'Living on the Aryan Side in Poland. Gender, Passing and the Nature of Resistance'. In *Women in the Holocaust*, edited by Professor Dalia Ofer and Lenore J. Weitzman, 187–222. New Haven, CT: Yale University Press.

7

'Trademark Wars'

Naxals versus Criminal Extortionists in India

Pavan Mamidi

One of the major means of funding for extralegal and rebel organizations is extortion of civilians, but it is attended by the challenge of informational asymmetries that exist between extortionists and their victims.[1] For extortion to succeed, threats have to be credible, and for threats to be credible, the extortionist's claims to membership of a threatening group have to be convincing. How can victims tell whether their extortionist truly represents an extralegal organization and has access to its means of violence? Signalling is an important way in which identities and threats of violence are made credible. However, under conditions in which signals are not fully discriminating, conditions afforded by the existence of multiple victim types, deceptive mimicry can occur. Criminals who do not belong to an extralegal organization can attempt to co-opt its intimidating reputation or trademark by 'passing themselves off' as its members, and extort money from civilians. As a result, extortionists who really do belong to an extralegal organization must rely on costly-to-mimic signals to overcome this problem. The idea of protection and licensing of an extralegal organization's trademark may be found in Gambetta (1993), in which the brand equity of the Sicilian mafia is subjected to a rational choice explanation. He reasons that members who are entitled to use the trademark of the mafia can profit from the image of fearsomeness that it triggers. Schelling (1960) and Ellsberg (1968) have discussed the conditions under which threats may be made more credible.

Recent work that applies signalling theory, and the conceptual framework of deceptive mimicry (Gambetta, 2005), to the context of civil wars include

[1] I want to thank the following people, without whom this work would have been impossible: Diego Gambetta and Michael Biggs; David Laitin, Macartan Humphreys, Jim Fearon, Sharon Barnhardt, Valeria Pizzini Gambetta, Abigail Barr, and Juan Masullo J. for their comments during the December 2009 workshop on Mimicry at Nuffield College, Oxford and the process of writing this chapter; Mr. C. Anjaneya Reddy (former IPS and senior intelligence officer, Govt. of Andhra Pradesh) for introducing me to various officers and subjects interviewed for the research; and numerous gatekeepers on the field.

Pavan Mamidi, *'Trademark Wars'*. In: *Fight, Flight, Mimic*. Edited by: Diego Gambetta and Thomas Hegghammer, Oxford University Press. © Pavan Mamidi (2024). DOI: 10.1093/9780191802454.003.0007

Habyarimana et al. (2009, ch. 3) who provide experimental evidence from Uganda on ethnic identifiability. Holmes (in this volume), describes the conditions and scenarios that go into identity signalling and mimicry in the ongoing Iraqi civil war, and Pizzini-Gambetta (in this volume), deals with deceptive mimicry and signalling in the context of political violence in Italy. Similarly, Hamill (in this volume) addresses issues of identity signalling and mimicry in the political conflict in Northern Ireland.

The Main Characters

In this chapter, I examine how this challenge is met by the Naxalites (or Naxals), a left-wing guerrilla group that operates in rural and semi-urban India.[2] They face the problem of authenticating their identity to their victims in the face of other players in the landscape who try to mimic the traits of the Naxals and also try to extort civilian victims. This study is based on Naxals and Naxal mimics operating in the Telangana region, a region that has only recently become a new state in India. Fearon (in this volume) develops a game theoretic signalling model of social mimicry around Gambetta's (2005) stylized model that includes the three players: the model, the mimic, and the dupe (the receiver of the signals). I use the same stylized model of Gambetta in identifying the main characters of this chapter (see Table 7.1).

The Naxals (the Model)

The Naxals as a rebel underground organization have been waging a protracted war against the state for the past four decades. The organization originally started in 1967 in a village called Naxalbari, a small town in West Bengal, as an armed movement with the aim of advancing a Communist revolution in India, and protecting the rights of peasants and lower-caste members of society. They continue to operate in tribal villages and forests of India (see Figure 7.1), and seek to provide governance services and other public goods, including dispute resolution services, in areas that are underrepresented by the formal state. They may be construed as political entrepreneurs who supply governance services in regions that have an unmet demand for these services. In turn, for the services that they provide, they tax

---

[2] The Naxalites are an armed group of militants, ideologically motivated by varying strands of Marxism, Leninism, and Maoism. They are usually members of the single largest, pan-Indian, banned underground party called the Communist Party of India (Maoist).

Table 7.1 The main players

Players	Role	Characteristics
A. State	'Taxes' civilians	Prefers to monopolize taxation and provision of governance services. Claims legitimacy
B. Naxals (Model)	'Taxes'/'Extorts' civilians	Prefers to monopolize taxation of civilians and provision of governance services. Claims legitimacy. Competes with state and Naxal mimics in collecting 'taxes'. They seek to ensure that the victims of their extortion can differentiate between them and the Naxal mimics
C. Naxal Mimics 1. Entrepreneurial criminals	Extort civilians	Pass off as Naxals, and muddy the signallings relationship between Naxals and civilian victims
2. Ex-Naxals	Extort civilians	Pass off as current Naxals and extort civilians They have most of the traits of currently active Naxals, except the organizational and network resources
3. Sympathizers	Extort civilians	Pass off as current Naxals and extort civilians They have several traits of currently active Naxals, but not all the costly cues of the Naxals
D. Civilian Victims (Dupe)	Pay money	Correctly identify Naxals and pay money, and correctly identify Naxal mimics and not pay money

civilians, which the formal state considers as extortion. In effect, the Naxals and the formal state compete for 'taxes' in payment for the competing public goods that they provide. Extortion is considered to be their main method of collecting 'tax', although other means such as running protection rackets and illegally selling timber and minerals have been reported. The Union Ministry of Home Affairs in India estimates that they raise close to INR1400 Crore (equivalent to $300 million US) nationwide annually. Today, under the consolidated Communist Party of India (Maoist), the Naxals are considered to be at the peak level of organization in their history. Current estimates suggest that there are as many as ten to fifteen thousand Maoist Naxals who operate in India (D'Souza and Routray 2010).

The victims of extortions in the Naxal and Naxal mimicry landscape are ordinary civilians, business entities, and businesses located on a continuum between core Naxal territories and the peripheries surrounding these territories. They can be pretty much anybody—middle class, lower middle

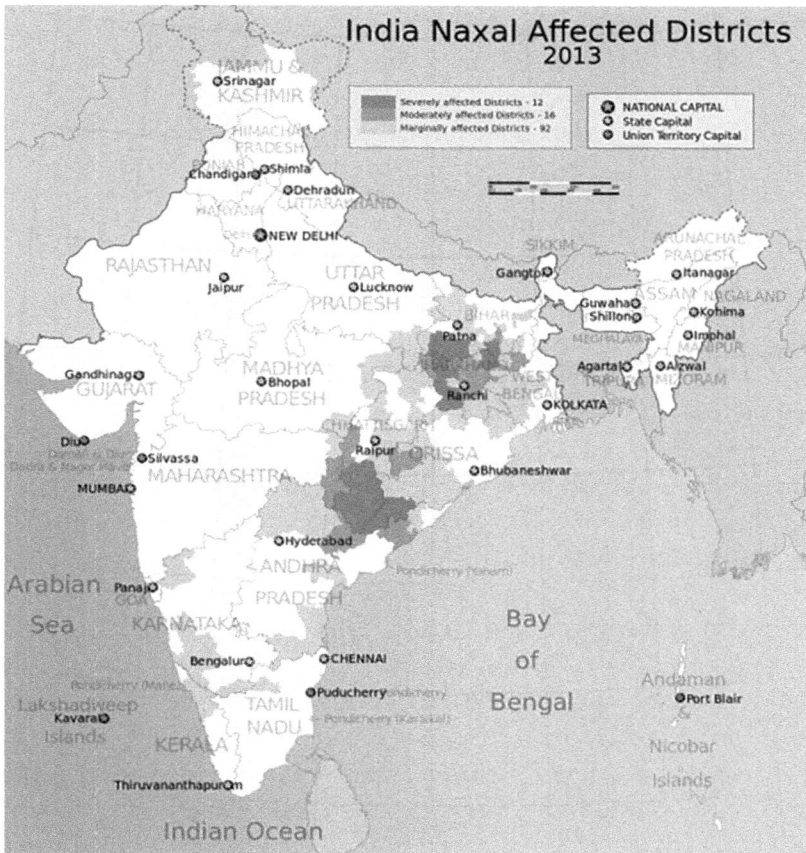

Figure 7.1 Areas affected by Naxalites in India

Source: Wikimedia Commons

class, salaried employees of factories and the government, wealthy individuals, owners of farms, business entities and businessmen, and government contractors. The Telangana region of India is comprised of some big cities at one extreme of a continuum (such as Hyderabad, Vijayawada, and Vizag), followed by mid-level district towns and small towns (such as Warangal, Khammam, and Adilabad), and much smaller towns and villages at the other end of the spectrum. Victims are typically residents of districts and villages, and are seldom located in the bigger cities. It is pertinent to note that some of these victims, especially those who operate as contractors for the government in extractive industries, such as mining, often have close connections with politicians of the ruling political party. When they get extorted excessively, their costs of production increase, and they are likely to leverage their connections with their political patrons, and trigger greater monitoring and conflict between the state and the Naxals through police action. Conflict between the Naxals and the state is largely an outcome of the competition between the Naxals and the state for the 'tax' that is imposed on the victims.

The Fake Naxals (the Mimics)

Naxal mimics are also called pseudo-Naxals, or fake Naxals, and typically operate in the geographic peripheries of core Naxal areas. They attempt to pass off as Naxals with civilian victims so that they can extort them. In other words, there are three competitors for coerced 'taxation' inflicted on civilians in these areas: the formal state, the Naxals, and the Naxal mimics.

Broadly, there are two types of mimics who operate. First, there are the criminally minded entrepreneurs who have absolutely no prior contact with either the Naxals or with any aboveground, politically oriented left parties. Second, we also have ex-Naxals and aboveground sympathizers, who were once activists in their own rights, but have now turned to extorting civilians under the banner of the Naxals.

Ex-Naxals are the most insidious type of mimics. They know how to behave like real Naxals and can pass off as Naxals better than other types of mimics who are ordinary criminals. The pool of ex-Naxals is typically comprised of two types—those who were arrested and served time and those who left the Naxal movement under a government-sponsored rehabilitation programme. Both of them are likely to have abnegated their primary means of survival, including family and community connections, or educational degrees, in order to demonstrate their commitment to the Naxals when joining them (see Mamidi 2010). They are people who 'burnt their bridges' with social and material resources to demonstrate their trustworthiness to join the Naxals.

This very specific feature of ex-Naxals, i.e. broken links with alternate means of survival that once helped them demonstrate their commitment to the Naxals and join the underground, can push them into a life of crime and extortion in their current lives. Many of them have been enticed by government rehabilitation programmes that give them money upon leaving the Naxal movement. But this money does not last long, and they soon realize that it is not enough to sustain a new life above ground.

As a result, many ex-Naxals are left with little else other than the skills of waging a guerrilla war and raising capital through extortion. They are people who have systematically and deliberately abnegated their resources in their pre-Naxal days that could have been presently useful in above ground society. With the exception of a few who originally joined the Naxals with portable technical skills, and who are now capable of hiding their tainted

Naxal history, the rest find it hard to get jobs and eke out a living. This is exactly when a transformation of their inner ideological motivations that they carried as Naxals is likely to occur. The original motivations of ex-Naxals can cease to exist and these ex-Naxals can become self-serving. They are likely to join the rat race of survival above ground by turning mimics.

Some of the mimics can also be Naxal sympathizers. Naxal sympathizers are typically above ground left activists, who provide support services to the Naxals. Although Naxal sympathizers do not abnegate alternate means of livelihood to the same degree as ex-Naxals, several of them join teams of Naxal mimics when socialist ideology disillusions them. In their roles as sympathizers they would have collected information about victims for the Naxals, and their abilities to do this are valuable to teams of Naxal mimics. As mimics, they cannot produce all the costly-to-mimic signals of ex-Naxals, but certainly are more convincing than ordinary criminals who become Naxal mimics. Since their marginal costs of producing Naxal signals are low too, they are enticed by possibilities of using these signals and making easy money from extortions.

Ex-Naxals already have the traits and qualities needed to pass off as currently active Naxals. Following Gambetta's (2005) schema, they may be described as 'ex-models' who become 'mimics', where the transformation from the former to the latter occurs at negligible marginal costs of producing signals needed to convince the dupes that they are the real models. In other words, while ex-models retain their outwardly visible qualities of models because it is almost costless to do so on the margins, their latent traits shift from ideologically and socially minded dispositions, to opportunistic and self-serving ones. This is an unusual type of social mimicry in which some of the real types stop being real, and transform to become fakes and challenge the existing real types.

Sources

Data for this research have been collected from interviews conducted in the Naxal-occupied Telangana region of Andhra Pradesh, south India, over a period between 2007 and 2009 and then again in some spells in 2010. Most of the data come from interviews with six categories of subjects: (i) Naxals (Ns), (ii) ex-Naxals (ExNs); (iii) Victims (Vs); (iv) Police (Ps); (v) Naxal sympathizers (Ss), and (vi) Civilians (Cs).[3] I was able to obtain access to

---

[3] Much of the data used in this research comes from the interviews conducted for my previous work on signalling and recruitment—see Mamidi (2010). I have extracted data from ten interviews with Naxals, fifteen from Ex-Naxals, fifteen from victims, including ten victims of successful mimicry, twenty from police personnel, and from scores of sympathizers and civilians.

various informants using personal contacts, including police officers, left-wing activists, and former university classmates who were activists during their youth. They were willing to provide introductions because we all grew up together in the Telangana region in the 1970s and the 1980s when being a socialist was much more common than it is today. Interview subjects trusted me with information because they saw me quite rightly as a 'harmless academic researcher' with no ulterior motive to cause any harm to them.

Ordinarily it is impossible to find victims of successful mimicry in everyday life. Victims of successful mimicry do not know that mimics have duped them. An opportunity to find victims of successful mimicry was provided when mimics arrested in failed mimicry attempts gave leads to victims they had successfully duped in their earlier mimicry attempts. They gave these leads to the police (possibly under duress), and the victims later confirmed that the same mimics in question duped them. I interviewed ten victims of successful mimicry, from leads provided by the mimics. I also interviewed the five victims of the unsuccessful attempts, and the five mimics. This allowed me to compare characteristics of victims who were duped successfully with victims who were not duped by the same Naxal mimics. I discovered that none of the victims that I spoke to were ashamed to speak about their misfortunes openly.

I am able to attribute several views and statements made by subjects to specific subject types (ex-Naxals, or Naxals, or Victims etc.) interviewed individually, indicated as codes in this chapter for the sake of preserving anonymity. However, much of my data was also drawn from informal group discussions that happened spontaneously under a tree or in a hut in the field, usually in the evenings over an evening meal or a cup of chai. In such situations, it was hard to record the identities of the specific subjects who may have made some of the noteworthy comments that inform this work.

Why Mimic-Extortionists Want to Identify as Naxals

Why do extortionists want to use the identity of the Naxals? Why do Naxals not disguise their Naxal membership altogether and approach victims as regular criminals? Is it not better for them to conceal their identities and preserve their reputation as ideologically motivated political activists? In fact, Elster (2004) points exactly to such a possibility. Hamill (this volume) and Pizzini (this volume) report instances of the IRA and of left-wing extremists in Italy passing as ordinary criminals respectively.

The main advantage of being seen as a Naxal (truly by real Naxals, or falsely by Naxal mimics) comes from the strong 'trademark' effect that the usage of the word 'Naxals' has on victims.[4] Much the same way that business trademarks reduce search costs of sources of value in markets, the name of the Naxals immediately creates a presumption of power and capabilities. The strength of their image makes threats more credible (most Ps, Ns, ExNs, Ss in the sample)[5]. Also, there are no known criminal gangs with comparable intimidating power that can extort in their own name in Telangana (most Ps). There is no evidence to suggest whether or not other comparable criminal groups in Telangana might have existed that were destroyed by the Naxals.

The Naxal trademark has been created over decades of repeated interactions between the Naxals and civilians. Media coverage of sensational events by Naxals over the years has contributed to building an aura around the Naxals. According to most civilians among the subjects interviewed in tribal villages, the Naxals represent the Robin Hoods of society benefiting the poor in some basic way; to others (especially the police) they represent undesirable, criminally minded terrorists plaguing the country and threatening to take it down the path of Communism (several Ps). However, according to most civilian interviewees and victims, they are people with a steely determination and high levels of endurance to pain, with the ability to live in forests under brutally hard conditions (several Cs and Vs). They are people who can accept high personal costs to follow through on commitments, including extortion threats (V1–4).

Using the Naxal trademark also conveys that a systematically organized group is backing up the threats of violence that accompany demands of extortion. When one set of Naxal members issues a threat, and for some reason is unable to carry it out, the Naxal organization immediately mobilizes other members to take over. Attempts to execute threats are relentless. In this respect there is a similarity between the Naxals and business corporations, because both follow the principle of perpetual succession of promises

---

[4] See Gambetta (1993) for application in the context of the Sicilian mafia.

[5] I will refer to my interview subjects by the following codes: (1) A Naxal subject will be referred to as 'N' followed by a number; (2) ex-Naxal as 'ExN' (3) victim as 'V' (4) police as 'P' (5) Naxal sympathizer as 'S', and (6) civilian as 'C' For reasons of security and privacy, I cannot provide names of subjects interviewed or the exact locations where they were interviewed (except, possibly the districts in Telangana). Several of these conversations are documented as raw visual and thematic maps made from memory, often several hours and even days after the interviews. They are multilayered with many iterations spanning multiple dates (between 2007 and 2009, and then again in 2010) and multiple locations. Many times it was not possible to clearly record the specific identities of subjects, especially when they were part of a larger group assembled for reasons other than for interviews. In such cases, I will refer to the group by a broad location where I encountered it.

and contracts through perpetual succession of organizational personhood. Irrespective of the individual agents who enter into agreements on the organization's behalf, the organization treats the obligations of the agreement as its own, even when those who initially formed the agreement have left. This assumption of 'organizational personhood', much like 'corporate personhood', likely reduces transaction costs in contracting with the external world by creating the presumption of organizational perpetuity. A road building contractor in Ranga Reddy District in Telengana [V8] who was once a victim of an extortion said: 'if you want to be a long-term business in the area, you need to be aware of the fact that a threat is not forgotten even if the issuer of the threat has been killed. The person who issues the threat is only incidental to the threat. It is a threat from an organization that outlives the people who form it'.

The presumption of succession of threat-execution by the Naxals is one of the most powerful features in their armoury. Small groups of organized criminals cannot match this feature given that they do not have nearly the same economies of scale that the Naxals enjoy in sustaining their brand and reputation for following through on threats. A sequence of failed attempts at following through on a threat may become increasingly costly for the Naxals as more and more members get caught. But given their relative size, they are better able to bear the costs of successive failures than smaller organized criminal groups, says a police subject in Hyderabad (P 5).

The expectation that Naxals will follow through on threats no matter what, complements the general civilian population's belief that the police cannot protect them against extortion threats. Civilians are convinced that a largely overstretched police cannot protect them. Civilians often consider the police as 'napunsaks' ('unmanly') (V3). This disparity in relative levels of certitude that civilians have about the two entities competing to tax the people—the Naxals and the government—creates the ideal conditions for extortion by parties using the name of the Naxals.

The Naxals indicate that the social purpose of their 'taxation' can mitigate the damage caused to their reputations by the use of coercion on their victims. A college student sympathizer in Adilabad (S15, Govt College) argues that 'some victims are less angry when they are extorted for an ostensibly "good cause" by Naxals than by criminals who do it purely for personal gain'. This thinking can tip victims' decisions in favour of not reporting Naxal threats to the police. This psychological position may also help victims reduce the cognitive dissonance of being extorted. The Naxals and Naxal mimics recognize the added benefit of using the Naxal trademark in psychologically steering their victims away from approaching the police.

Evidence that Naxal Mimicry Exists

Although there is no centralized record of instances of thwarted Naxal mimicry in the country, Naxal mimicry makes regular news in Naxal states. But how do we know that any of these arrested extortionists that the police claim to be Naxal mimics are really Naxal mimics? It is a distinct possibility that the police as an organization try to underestimate the number of Naxals in the state in order to appear that they have done a good job eliminating the Naxal problem. In theory, the police organization can try to pass off captured Naxals as Naxal mimics to trivialize the problem of Naxals, and thereby improve their image among the civilian population.

We can address this particular doubt by looking at the incentives that police officers have as individuals, which are likely to be different from incentives of the police organization as a whole. They are set in such a way that individually they would much rather claim that they have arrested real Naxals than fake ones. There are more rewards for arresting real Naxals than petty criminals trying to extort in the name of Naxals. Police officers get rewarded with promotions when they succeed at capturing Naxals and not when they capture petty Naxal mimics. This means that there might even be incentives for them to try and pass off Naxal mimics in their custody as real Naxals. Allegedly this is a common practice. There are numerous complaints by rural civilians that police officers capture innocent civilians and charge them as Naxals. Interviews in Nallamalla tribal villages indicate that there are several farmers and tribal folk who have been taken away by the police or even shot in 'fake encounter' cases (this is information from a group discussion with Cs in Nallamalla village). If at all, this trend confirms that individual police officers indeed have incentives set in such a way that reports they make of arrests of Naxal mimics are only likely to be an underestimate of the real figure.

But there is another possible set of doubts that we must reckon with before we take the existence of Naxal mimics for granted. First, there is a real possibility that the police officers do not have a concrete way of distinguishing between real Naxals and Naxal mimics. Allegedly they torture their captives to get confessions. But confessions obtained from torture are not reliable given that Naxal mimics are likely to accept being real Naxals under duress. Second, when real Naxals get arrested, they can claim to be Naxal mimics and not real Naxals in order to get a shorter jail term. Being a Naxal mimic is less of a crime than being a Naxal according to the penal code. It is theoretically conceivable that Naxal mimics are a mere fiction invented by captured Naxals in order to escape higher levels of incarceration. Captured individuals claiming to be Naxal mimics can potentially be real Naxals passing off as Naxal mimics. This is not hard to imagine because Naxals would know exactly what cues and markers they need to conceal when the police capture them. Elster (2004) proposes that politically oriented rebel groups may sometimes mimic ordinary criminals when they engage in extortion in order to protect the image of their group. Hamill (in this volume) reports a similar phenomenon with the IRA, whereby members sometimes engage in extorting money as ordinary criminals.

Let me briefly analyse the incentives of real Naxals to ascertain the viability of the above-mentioned doubts. Interviews with currently active Naxals reveal that Naxals are dead against the existence of Naxal mimics in the landscape (Most Ns in the subject pool). They feel that any news of the discovery, or incarceration of Naxal mimics has a pronounced effect on the value of their threats among civilians. Therefore, they are intensely averse to any news about Naxal mimics making it to the public attention via the media. Their animosity towards Naxal mimics is evidenced by the extraordinary lengths they are willing to go to capture them, including, but not restricted to, tracking them down in villages and killing them.[6] They issue standing instructions to members not to declare that they are Naxal mimics if they are arrested. They are worried that there is a risk that it will make news and that it can undermine their trademark in the marketplace of extortions. In fact, a Naxalite (N2) of a senior standing reports that they try to promote a code of honour among their members to uphold their identities with pride, even if it costs them in the short run. It is reported that they may even send incriminating evidence to police that a captured Naxal is a real Naxal if the captured Naxal attempts to pass off as a Naxal mimic.

It is also well worth considering that Naxals who are arrested and await trial as real Naxals, can receive full legal support from above-ground civil rights lawyers who are allegedly funded by the Naxals. This support likely reduces their sentence significantly. If arrested Naxal members claim to be mimics against the Naxal organization's wishes, the group likely withdraws legal support. Withdrawal of legal support greatly weakens the position of the Naxals undergoing trial. Given the low success rates of state-appointed legal defence in India, arrested Naxals likely prefer to incur charges as real Naxals having some legal support, than incur charges as Naxal mimics without any.[7]

---

[6] Sympathizers in Warangal district report instances in which Naxals have informally learned of victims of extortion through their networks, and approached them for details of the Naxalite mimics.

[7] This is a fact that is sometimes missed by victims when they make estimates of the identities of their extortionists on the basis of the price that the extortionists set when extorting. They are likely to overestimate the likelihood that the extortionists are Naxal mimics when the price is low. However, in reality, genuine Naxals prefer setting a high price to reduce their overall costs, which includes the costs of legal redress along with costs of incarceration.

There is more conclusive evidence of the existence of Naxal mimics. First, both ex-Naxals as well as the Naxals themselves (interview subjects from Khammam, Warangal, and Ranga Reddy districts) unequivocally admit to the existence of mimics. They also confirm that they face an ongoing challenge of preserving the credibility of their threats in rural Telangana, and the fact that Naxal mimics increase their costs. Second, there is also evidence in the type of weapons that Naxal mimics possess when they are captured during extortions. Naxal mimics do not usually have real weapons (even ex-Naxals surrender their weapons), and tend to rely on fake guns and grenades which can become a give-away if victims realize, as they often do, that extortionists are brandishing toy weapons. In contrast, arrested Naxals are found with real weapons. They would much rather carry real weapons because they can better protect themselves against possibilities of capture during extortions with real weapons.[8]

Forms of Signalling

Third-party Authentication as Costly Signalling

Naxals often use the method of third-party authentication to signal their identities. The use of third-party references may be considered as a form of costly signalling, insofar as only genuine Naxals are able to leverage well-known third parties to authenticate their identities. Consider the following case involving the use of third party references: A victim (V7), who is a real estate developer in Medak district close to Hyderabad, receives a threat by phone from a caller who claims to belong to a local Maoist *dalam* demanding a sum of '25 lakh (approximately $50,000 US), deliverable at an appointed place and an appointed time. The caller threatens that V7 will be killed if he does not comply. Also, the caller indicates that V7 can confirm the authenticity of the caller's identity by approaching a noted above-ground Naxal sympathizer and artist (who is well known in the Telengana region of Andhra Pradesh), and ask him about the genuineness of the caller's identity. The caller also indicates that he will telephone this famous above-ground sympathizer and confirm that he made the specific demand from V7. Soon after, V7 meets the famous artist, who then confirms that the threat-maker is indeed a genuine Maoist guerrilla. V7 complies with the demand immediately.

The above example illustrates that the use of a third party to authenticate the claims of a Naxal extortionist requires the availability of a third party whose position and knowledge are well known in society. The accreditation

---

[8] Even if some of the disobedient types want to insure themselves against higher incarceration charges.

service in itself is a high cost resource that genuine Naxals can afford and mimics cannot. In the above case, the third party turns out to be a noted left-winger, with a reputation for deep linkages with the Naxals. But there is a serious problem with this method. Even if there were several noted above-ground left-wingers providing services of authentication in circulation, there is likely a limit on how many extortions they can abet by providing authentication without the above-ground authenticator inviting the attention of the police. The authenticator can of course always get away from legal action against them by making disclaimers about his role in abetting the crime. Police admit that an above-ground authenticator can claim to the authorities that he was merely authenticating the veracity of a threat against the victim in the interest of the victim (most Ps).

In theory it is also possible to conceive of less known third-party authenticators who can bridge information gaps between Naxals and victims. However, victims are not likely to believe in these less known third-party authenticators. Less well-known third-parties make for weaker authenticating signals. Given the added transaction costs of forming trust in less known third-party authenticators, victims are likely to give more credence to direct contact with their extortionists than go through people they barely know.

Repeat Transactions

A second strategy that Naxals sometimes use to solve their authentication problems is to focus on victims who can be extorted multiple times rather than on a one-off basis. This strategy is neither incommensurate with costly signalling, nor an alternative to it.[9] It is merely a method that reduces the need for costly signalling on a repetitive basis. Both the Naxals and their victims have to incur the costs of authentication by signalling only once, i.e. at the start of their relationship, and not thereafter. So, if at some point in history, the Naxals strike enough relationships with repeat victims to sustain their operations, we should expect to see them not approaching new victims and not encountering the problem of authentication anymore. We should also expect not to see Naxal mimics in the landscape given that civilians, other than the existing repeat victims, will stop receiving (or expecting) threats.[10]

---

[9] Bacharach and Gambetta (2003) talk about signalling quality by signalling identity. It suffices to send a persuasive signal that you are i to imply k.

[10] Smith and Varese (2001) propose a game theoretic model of extortions by mafiosi, involving criminal mimics who extort entrepreneurs, and conclude that repeated violence is an essential part of maintaining the Mafiosi trademark.

Again in theory, repeated extortion from victims does not reduce the importance of signalling in extortion for the Naxals. Further, it is not as if there are no costs associated with extorting repeat victims in comparison. There are fluctuations in the arrangements that the Naxals have with their repeat victims. Often, income streams of victims change, varying the victims' abilities to pay. The Naxals require continuous efforts to verify victims' stories of changing fortunes. There are costs of continuous renegotiation demanded in such situations, leading to irreconcilable misunderstandings, and pushing the Naxals into positions where they deem it necessary to carry out their threats of punishment. The Naxals run the risk of their credibility being diminished severely if they do not carry out their threats on occasions of defaulted payments.

There is also the likelihood that repeat victims change their assessment of the benefits of being free from threats over time (see Shavell 1993). When victims become more desirous of being left alone, they become more willing to default on their payments.[11] Again, this is likely to push the Naxals into a position where they have to demonstrate the seriousness of their threats, exposing them to the risk of losing their credibility among victims.

It is also well worth considering that the Naxals comprise mobile bands of members, who are never permanently stationed in one geographic area. This makes them incapable of sending the very same members recognized by repeat victims to make collections. There are reports that some Naxals solve this problem by using secret password phrases that they share with repeat victims and that signify that the extortionist is a genuine Naxal. But this system, which works somewhat like using passwords to draw money from ATM machines (the repeat victims are the ATM machines!), runs into problems when password phrases leak and are misused by Naxal mimics. There is a story told by a police officer (P11) of an employee of a cement factory in Adilabad district who became privy to one such secret password phrase that the Naxals used in collecting money on a regular basis from the factory. The employee then secretly formed his own little Naxal mimic operation and sent his band of mimics to collect money from the factory in the name of the Naxals.

But primary among all reasons there is one that compels the Naxals to extort non-repeat victims and therefore inevitably depend on continued signalling of their authenticity. They often have expansionary plans that are likely to make the present value of large, one-off payments more appealing

---

[11] Shavell (1993) proposes conditions under which extortionists might find it costless to attempt repeat extortions, but also brings out reasons why they raise transaction costs.

to them than a stream of repeated payments over time. Repeat extortions do give a steady income, but this income may not be enough to satisfy spikes in expenditures. This is especially true when the Naxals need to make large purchases, or incur large fixed costs for carrying out new initiatives, such as setting up new manufacturing units, or moving into new areas.

Given that signalling for extortions is an unavoidable means, it is also important to realize that the process is not entirely a costless one either. Although the marginal costs of producing signals and cues for the Naxals to credibly tell their victims that they are indeed the real deal are low, there are significant cumulative costs imposed by the presence of Naxal mimics who muddy the communication between the Naxals and their victims. The constant need to invent new signals when their old ones are copied and become useless adds to the costs of bridging informational asymmetries with their victims. Evidence of creating new signals comes in the form of the Naxals needing to educate victims about some of their lesser-known traits from time to time. For instance, a Naxal reveals that when his *dalam* (platoon) in a particular region in Adilabad discovered that more and more mimics were adopting a calm and polite conversational style (a typical Naxal style that prevailed for a while) when threatening victims, the Naxals decided to make it more well known that the Naxals were not merely about a particular conversational style, but also about substance derived from intense social learning and understanding (N7). The Naxals in this case could accomplish doing this by using their network of above-ground sympathizers with extensive contacts with civilians in propagating the view that it is only the Naxals who have an ability to use a certain kind of 'social logic' that would become apparent when they use it, and that they alone can use it (corroborated by ExN4 and a S5 in Adilabad).

Costly to Mimic Signals Versus Cheap Talk

Analysis of data indicates that the typical cues and signals used by the Naxals and the Naxal mimics for extortion lie on a continuum ranging from low-cost appurtenances, such as clothing, stationary used for the issue of threats, physical appearance (such as Marxist beards and long hair), to the more costly to mimic actions—such as the idiomatic usage of Marxist or leftist language, brief acts of self-harm such as drawing blood from the thumb, pre-emptive acts of violence, such as firing guns into the air, and strategic use of personal information about victims (see Table 7.2). There are strong reasons to believe that variation in the costs of mimicking the cues and signals produces variation in the levels of credibility attached to them.

Table 7.2 The signals

Signal	Costly to Mimic by ordinary criminals?	Costly to Mimic by ex-Naxalites?	Response of Victims/Notes
Clothes Military fatigues, shoes, sandals, slippers, khadi, kurthas, jholas	No	No	When demanded amounts are very low, uninformed, risk-averse victims oblige
Physical Appearance Long hair, flowing Marxist beards, fitness	No	No	When demanded amounts are very low, uninformed, risk-averse victims oblige
Printed matter Pamphlets, letterheads, books, etc.	No	No	When demanded amounts are very low, uninformed, risk-averse victims oblige
Real weapons with peaceful demonstration	Yes	Not necessarily	Carry out the demands
Fake weapons, air pistols, plastic guns and grenades	No	No	Even though these are easy to acquire, they can influence victims significantly. Victims may not be able to ask for demonstrations because they fear that by doing so, they might invite violence from an extortionist who is carrying real weapons. However, a perfectly informed and rational victim would know that real Naxalites would find peaceful ways of demonstrating that the weapons are real and not hurt the victims
Acting skills Name-dropping, consistency, role-playing, recitation of poetry in dramatic manner, speech giving;	Yes	No	Ex-Naxalites who turn mimics can use this successfully. But ordinary criminals cannot afford this signal
Cognitive/intellectual skills Knowledge of Marxist literature, extempore recitation of poetry showing internalization of poetry, knowledge of politics, government policies, quality of language used—whether embellished with sanskritized words.	Yes	No	Ex-Naxalites who turn mimics can use this successfully. But ordinary criminals cannot afford this signal

Continued

Table 7.2 *Continued*

Signal	Costly to Mimic by ordinary criminals?	Costly to Mimic by ex-Naxalites?	Response of Victims/Notes
Composition of Group Women in group who carry weapons, speak, and with authoritative confidence; lower-caste cues and signals, such as darker skin tones, multi-person display of skills.	Yes	Yes	Even ex-Naxalites cannot ordinarily put together a team, especially with women. It is very hard for ex-Naxalites to organize mimicry teams because they have higher levels of distrust among each other. They may fear that potential ex-Naxalite partners may still have links with real Naxalites and give them away
Self-harm Drawing blood from thumb to anoint victims' foreheads	Yes	No	Ex-Naxalites who turn mimics can use this successfully. But ordinary criminals cannot afford this signal
Insider knowledge Precision and reasonableness of money demanded, with knowledge of private information of victims	Yes	Yes	Even ex-Naxalites cannot gain access to private information about victims over a wide geographic area

Keeping all else equal, including victim types, costly to mimic signals tend to produce greater levels of credibility than do low cost signals. Also, as one expects, when an extortionist is a genuine Naxal, his signals are likely to be costly to mimic. A Naxal subject N3 in Ranga Reddy district reveals that Naxal extortions involving one-off, first-time victims are likely to be occasions when Naxals deliberately signal their attributes using costly signals. Even if they arrive at the victim's place in their military fatigues, sporting their long, flowing Marxist beards, carrying messages on Maoist letterheads, they are acutely aware that these highly imitable attributes do not necessarily sway the beliefs of their victims their way. They recognize that in the long run, they have to depend on more substantive bases to convince their victims.

I start with low-cost signals. Although they are available to both the Naxals and the Naxal mimics alike, the Naxals rely less on them than do mimics. So, let us focus on how Naxal mimics use them, and also manage to succeed with them. One discovers that the Naxal landscape is replete with instances of successful mimicry involving low-cost signals. Police indicate that most of these instances do not involve anything more elaborate than inexpensive 'Naxal types' of clothing, such as *khadi kurthas*, red bandanas, or military fatigues.[12] The mimics wear canvas shoes or sandals (associated with military or leftists), and deliver messages on fake Maoist letterheads, with smears of red creating impressions of thumbprints in blood. They also sometimes make phone calls from area code locations that match where they say they are calling from. It appears that some of these low-cost signalling strategies may have worked for a temporary period of time in regions that were relatively new to the Naxal phenomena, where civilian victims were willing to wrongly infer that mimics are Naxals (several Ss, Ps across Telangana).

Observing data obtained from victims of successful extortions by Naxal mimics, we notice that the amounts that are demanded cannot hurt the victims terribly (several Ps). Several police interviewees suggest that demands are low and within manageable means of the victims. Sometimes the demands for deferred payments are accompanied by indiscriminate grabbing of what is immediately available in victims' households—television sets, cooking utensils, and even curtains and tablecloths. There are also instances of mimics accosting women in the households they target. This form of petty grabbing and misdemeanour is strong evidence of petty criminals and not Naxals. The Naxals would never forcibly steal household items. Even if they are valuable (such as television sets), they are cumbersome to carry and hard to monetize. Secondly, Naxals have strong norms of gender equality that prevent them from teasing women. This is a view confirmed by Naxal subjects, who are always eager to talk about their equal treatment of women.

Most of the victims successfully extorted by Naxal mimics admit that they did not think enough about the magnitude of demands it would take real Naxals to risk being captured by the police. They simply have no clue about how much the Naxals must extort to make it worth their while to take the risk. These victims are oblivious to any strategic thinking about the extortionist's costs and benefits from engaging in such extortions.

Victims of Naxal mimics who used low-cost signals also express a complete lack of faith in the police's ability to save them from extortionists and do not report threats to the police—although, not surprisingly, most of these victims appear to be an educated, steadily employed, salaried, non-adventurous, and law-abiding type of people. I conjecture that these law-abiding traits render these victims excessively susceptible to the fear of punishment for breaking any kind of law or rules set by authority—the very same traits that perhaps also make them susceptible to the threats of the Naxal mimics.

---

[12] Khadi is a rough, homespun cloth associated with villages, Gandhi, and anti-industrialization.

In contrast to the victims who fell prey to Naxal mimics and do not report to the police, those who do not fall prey, but report to the police, come across as being of exactly the opposite type—the types who are likely to be less law-abiding than the ones who do not report. They are the local 'Daarkaaris' ('goons'), non-salaried, self-confident, and reckless types. It appears that most of them are likely to be construction contractors with a reputation for getting things done even if it means cutting corners. They sound fairly irreverent of the police, but concede that they would tell the police of a threat. This is not because they expect the police to protect them from the extortionists as such, but because if confronted by mimics it is nearly costless to solicit police support for whatever it is worth.

In contrast to Naxal mimics who extort with a repertoire of cheap talk, the Naxals insist on being authentic. Parts of their presentations are deliberately geared to be costly to mimic. Even though Naxals do not always talk in an eloquent, ideologically charged manner in private, they choose to do so when they address victims. Occasions of contact with victims are exploited thoroughly—a Naxal subject in Adilabad indicates that they are treated as opportunities to generate and sustain the sensational persona on which their strategy of extortion depends. When they spot any gaps between how they sometimes behave in front of their victims and the public persona they have cultivated in the minds of the people, they spare no efforts in filling them with drama and show. Grand speeches are made, revolutionary verses are recited extempore, famous scholars are quoted, and government policies are denounced using complex and erudite reasoning. Victims often find themselves in the unenviable position of being rudely woken up in the middle of the night, and made hostage audiences to unfolding Naxal tirades. These acts by Naxals are designed to not only reliably signal their identities but also to leave a strong impression in the minds of the victims long after they have left. They want to use every opportunity to add to their pre-existing public persona (Group discussion with Ns, Ss, and ExNs in Adilabad).

The felicity and ease with which they perform their cognitive and artistic acts are critical to the Naxals' success at extortion. Forgotten lines in recitations of verses, hesitations, and stammering and stuttering in speeches are considered failures. Bad performances show a lack of authenticity in the persona. As a result, they make a special effort to commission their star thespians to undertake initial meetings with victims of extortion. When such shows end, they go home and gloat about how well they played their roles, or how convincing they were in scaring the living daylights out of their victims (N4, Adilabad).

Although extensive acts of self-harm are normally not a part of ex-ortion related signalling, minor forms do sometimes occur. Some ex-Naxals from Khammam district report that in their days as Naxals, they would cut incisions in their thumbs, and draw blood to smear on the foreheads of their victims (ExN 2–5). They explain that such acts have a highly compelling effect on victims. Interestingly, an ex-Naxal, who is now a medical doctor and professor (ExN in Hyderabad), reports that his *dalam* members engaged in subterfuge sometimes. They secretly dabbed a bit of local anesthetic (of the type that is used by orthodontists) on their thumbs before they went out for a visit to a new victim. This minimized the pain of performing pseudo-religious rituals of slicing thumbs and anointing victims' foreheads. Clearly they were not averse to minimizing the costs of signalling without compromising on the quality of the performance.

Carrying weapons, including impressive-looking Kalshnikovs, or hand grenades is also a normal part of the show that Naxals put up for their victim's benefit. Firing in the air and lobbing grenades in the yards of village landlords are reported as fairly common occurrences. Pre-emptive blasting of rich farmers' cowsheds and electric pumps are also reported (ExN, Khammam). Although weapons can be used for immediate extortions of wealth or jewellery that victims store in their homes, the Naxals realize that extortions based on grabbing what is immediately available can only lead to fewer and fewer civilians keeping valuables or money at home in the long run. They would eventually have to go back and depend on extortions based on deferred payments. The Naxals admit that they carry weapons to first meetings with victims for show rather than immediate extortions, although they do not discount the possibility of using them for self-protection from police capture on such occasions.

There are also some Naxal signals that are dependent on the social composition of the team that makes the first visit to a victim. Some of these teams deliberately include armed women. Inclusion of women in extortion teams is one of the most convincing signals that can be used by the Naxals. The Naxals are among the only extra-legal entities in India that have a significant percentage of women in their ranks. They are able to induct women in their cadres by employing a gender sensitive, egalitarian rhetoric that is typical to Communist groups (Group ExNs, Khammam).

The Naxals explain that when women carrying weapons, speaking with an independent voice, and exhibiting strong intellectual skills arrive at a victim's house, the victims are almost certain to conclude that they are real Naxals. The effectiveness of this signal can be explained by the fact that India is a highly gender-biased society, and women in rural areas are bound by predetermined roles in households or farms, with relatively limited mobility outside their family boundaries. They face added social costs for associating with men outside their families. As a result, their participation in organized crime is likely to be far less than men. It is almost impossible for ordinary criminals to enlist the services of women to become partners in crimes like extortion.

It is also found that Naxals also deliberately send groups rather than individuals in their first visit to victims. They like to show that everyone in the group can exhibit the traits of a Naxal (which as we shall shortly see the ex-Naxal Mimics cannot replicate). The likelihood that all members of a group can display costly to mimic signals is higher for the Naxals than for an ordinary group of criminals mimicking the Naxals. In other words, a single genuine Naxal may find it harder to separate himself from a single mimic, than a group of genuine Naxals to separate from a group of mimics. Putting together a group of convincing mimics is more costly. Further, signallers can diversify on the types of skills or signals they deliver. Some can give speeches, some naturally look angry, some look 'ugly and scary', and some others talk calmly and softly. There is some evidence that they also deliberately choose members with 'dark' skin tones to keep up the lower-caste image of the Naxals. These strategies indicate that Naxals frequently enact an extreme version of their own public stereotypes in order to successfully convince victims (Group Ns, ExNs, Ss Khammam).

But we will see that some of these signals are effective not merely because they are costly to mimic. The circumstances in which these signals are used also determine their effectiveness. Differences in types of victims and extortionists, quantities of money demanded, and relative costs of incarceration produce different success rates of extortions for the same signals. Given the wide range of possibilities in the combinations of these variables, and given that the signalling game is a continuously evolving and a dynamic process involving two competing entities, the strategy of the Naxals remains focused on optimizing returns on their signals and employing context-specific differentiating features that set them apart from Naxal mimics.

Let me consider how the receptivity of victims impacts the effectiveness of costly to mimic signals. The Naxals' challenge of signalling their identities to their victims is often limited by how much their victims already know about true Naxals, and how much they can recognize costly to mimic signals as traits of real Naxals. The Naxals have to operate within a narrow margin of differences that potential victims see between real Naxals and the fake ones.

Consider the challenge that an ex-Naxal (ExN4), faced when trying to pass off as a Naxal on his first mission of extortion in the Bongir area of Nalagonda district. ExN4 makes an unannounced visit to a local rice mill owner's home in the middle of the night to make a demand. For some unknown reason he is unaccompanied by a team. He introduces himself as a member of the famous Rachakonda *dalam* that operates in the local area and explains why he is there. He sits down with the mill owner at his dining table, and briefly communicates a proposal from the Naxals asking him to send a 'donation' of INR 15 Lakh (approximately $30,000 US) by a certain date, stating the reasons why it is important for the mill owner to comply with the request.

Much to his surprise, ExN4 finds the mill owner very calm and composed during the entire interaction. He is not in the least flustered by ExN4's visit. In fact, he is amenable to a reasoned and calm conversation. The mill owner agrees in principle to send the money to the *dalam* but explains that he would need more evidence from ExN4 (or someone else from the *dalam*) that he really is a Naxal and acting on their behalf. He mentions that there has been a series of extortions of petrol stations by Naxal mimics in the neighbourhood, and that he is not sure whether he can rely on ExN4's word that he represents the real Naxals. Sensing the scepticism of his host, and not knowing any better to do at the time, ExN4 frantically attempts to demonstrate his Naxal identity with even greater emphasis. His voice grows louder as he expounds his intellectual and ideological inclinations, including his resentment of the exploitation of the poor. He even tries to steer the discussion toward topics of Marxism to show how well-read he is in revolutionary literature. ExN4 hopes that the mill owner can recognize that ordinary criminals are not as articulate as he is.

The mill owner continuously appeals to ExN4 to be reasonable and understand his uncertainty. He stresses that if ExN4 were a real Naxal he would understand his predicament and not take it personally (much like how airport security people tell irate passengers they frisk at security gates that there is nothing personal in singling them out for the frisking, and that it is all in their own personal interest to cooperate). This discussion goes on for half the night but ExN4 is unable to convince the mill owner that he is a Naxal.

It slowly dawns on ExN4 that he is using a futile strategy to convince his victim. It turns out that the mill owner is a 'non-mulki' (non-local) immigrant from a different state. While he knew that Naxals were 'morally' oriented, he did not know anything more about Naxal ideology, leave alone names of thinkers and visionaries in Communist literature. ExN4 returns to the camp and explains what happened. The *dalam* (platoon) members carefully evaluate what went wrong with the interaction and decide not to punish the mill owner. They agree that the problem was with their communication with the victim and not the victim.

The above example illustrates an important point about signalling—that the inferences that receivers draw about unobservable traits of signallers in signalling interactions is premised on the assumption that receivers are familiar with the connection between costly to mimic signals and hidden traits of the signaller. Even if there is a necessary connection between the signal and the trait that warrants a causal inference of the latter from the former, and the receiver of the signal is oblivious to the connection, or is unable to make the connection, the value of the costly to mimic signal is lost on the receiver.

The challenge of operating within the narrow margin of differences between real Naxals and the Naxal mimics in the perception of victims is further complicated by the fact that many Naxal mimics are ex-Naxals and Naxal sympathizers who have had prior exposure to the Naxals and turn to extortion as a way of livelihood. Ex-Naxals make especially good Naxal mimics because they, too, have almost all the needed costly to mimic signals that Naxals have. They paid their dues in acquiring these markers when they were Naxals. They, too know the idiom of Marxism, can walk the walk, and talk the talk of the Naxals. Similarly, above-ground Naxal sympathizers, too, know something about the Naxals and can pass themselves off as Naxals for purposes of extortion.

Given the presence of ex-Naxals and Naxal sympathizers, many of the costly to mimic signals of the Naxals that I considered so far become somewhat questionable in the minds of potential victims in certain areas of the extortion landscape. Take for instance, the cognitive and intellectual skills that Naxals use in their showy demonstrations to victims. The ex-Naxals turned mimics also know how to spout revolutionary poetry, give speeches, and talk Marxist shop. They, too, can speak emotionally, sound self-righteously angry, and draw blood from their thumbs. The drama of revolutionaries enacted in front of victims of extortion is not new to them.

Consider the example of an ex-Naxal (ExN6 in Ranga Reddy district), recently caught for trying to extort as a Naxal mimic. We can see how his experiences as a Naxal helped him fine-tune his presentation skills as an extortionist. ExN6 is known for fabricating Kalashnikov-look-alike assault weapons for use in extortions. He admits that he acquired the prior experience in repairing weapons and using machine tools to fabricate them when he was a Naxal. He narrates the interesting story of one specific occasion when he was still a Naxal. His *dalam* learned that some local Naxal mimics were using fake plastic toy pistols and Kalashnikovs to fool their victims. They also learned that victims in the areas had recently become more sceptical about threats issued by Naxals. News about the mimicry practices had spread around in Khammam and triggered greater caution among the victims. The *dalam* noticed an unprecedented collateral phenomenon—that civilians, too, started to acquire fake guns and keep them at home for show in case they were threatened by Naxal mimics. This is something that the subject's *dalam* had never seen before in their entire Naxal careers.

In response to this move by the victims, ExN6 explains, the *dalam* deliberately started to use a new tactic (i.e. the tactic of firing in the air with their guns whenever they approached their victims in that area). This was to demonstrate that they were carrying real weapons. ExN6 says that, finally, when he turned a Naxal mimic, the only way he could match the image of the real Naxals was by carrying a weapon he could fire in the air and make his threats appear menacing and credible. He admits with pride that this specific tactic worked with many victims until he was caught. The Naxals apparently got wind of him extorting people in the neighbourhood and sent word of his operations to the police. It turned out that the Naxals were unable to trace him, and because the government had the most recent knowledge of his whereabouts, they preferred to let the police catch him.

Signalling with Better Information and Price

There is some evidence to suggest that shifts in signalling equilibria are usually initiated when mimics come up with new techniques of mimicking the Naxals, disturbing a pre-existing equilibrium. When a fresh wave of mimicry starts, a percentage of the mimics get caught and make news locally. Thereupon, victims of threats get wary and approach the police instead of complying with demands of extortionists. At this stage all the players readjust their strategies. Also, when there is a significant time gap after a public announcement of a major mimicry-based extortion, the local civilian population is likely to forget lessons of caution learned from their previous experiences and go back to their original positions of vulnerability. Equilibria are continuously shifting phenomena in the extortion landscape.

Signalling relationships between Naxals and victims in different geographic areas are likely to vary with different levels of victim awareness. This is because information about extortions does not travel smoothly from one area to another and victims do not have opportunities to adjust their behaviours in short spans of time. Local areas where extortions have recently occurred and have been reported by the media witness localized production of risk aversion and availability biases among victims.[13] This might explain why some areas continue to witness only Naxal extortions using relatively costly to mimic signals, while some other areas witness extortions by Naxal mimics using low cost signals, and yet other areas witness ex-Naxal mimics using some costly to mimic signals of the Naxals.

Although there is insufficient data, I expect that victims in areas where mimics use low-cost signals are likely to be different from victims in areas where Naxals use costly to mimic signals; similarly, victims in areas where Naxals use costly to mimic signals are likely to be different from victims in areas where ex-Naxals use signals that are costly to mimic by ordinary criminals. Low-cost signals are not likely to be used where costly to mimic signals are in operation. In other words some signal types likely preclude the existence of other signal types in the same geographic area. However, I expect that different signal types can co-exist in the larger landscape of extortion where information does not flow efficiently between geographic sub parts.

Even though there is likely a continuously shifting scenario of types of signals that are used by the players in different areas of the landscape, it appears that the Naxals are armed with one particular signal that cannot be satisfactorily mimicked by others, not even the ex-Naxal mimics. This signal includes the relative precision with which Naxals can set their demands. It is also comprised of cues that the Naxals know some private things about the victim's financial or social situation, and that someone without extensive organizational networks could not have obtained such private information about the victims.

Consider the following case: A particular *dalam*'s members (N6,7) approach a local excise officer's bungalow to make a demand for a contribution of INR10 Lakh (approximately $20,000 US) to the local party fund of the Maoists. Among the many signals and cues that the group gives out, they also demonstrate that they know details of all personal transactions that the officer has recently undertaken in buying and selling land in the ongoing land boom. They recite the entire list of transactions, including details about the splits between the amounts that were received in cheques and the amounts that were received in cash. They shock him the most when they tell him that they even know where he keeps his bank locker keys in his *beer-vai* (chest of drawers) in his bedroom. This is information that the domestic servants in the excise officer's household have passed on to the Naxals. Until that point, the victim is begging and pleading to be left alone, falsely claiming

---

[13] See Gambetta (2005) for a discussion on production of availability biases by terrorist attacks.

that he does not have the money that the extortionists demand. But when the extortionists reveal their detailed acquaintance with the victim's financial situation, the victim is flummoxed, and convinced that the extortionists cannot be ordinary criminals. Surely, they must have the connections and the organizational capacities to look into other people's lives even though they are not from proximate social groups.

The Naxals are likely to acquire this particularly costly to mimic signalling capacity by drawing information about their victims from their extensive social networks of above-ground sympathizers (several Ps in the sample). They strive to collect details: whenever a landlord sells a piece of land, or a newly married upper-caste Babu receives a dowry, or a public servant receives a bribe, or a farmer has a good harvest. They pay special attention to new contracts in public works, and secretly obtain information about the values of the contract considerations. They endeavour to get information about sales and revenues of shopkeepers and local merchants, and find out when they are likely to make deposits in banks. They also make efforts to get information about wages given to workers in factories and the profit margins left for factory owners. They even try to get information about how much jewellery the local district collector's wife wears at weddings (Ss in Adilabad). It appears that they are better at getting information about civilians than even the government. In turn, there are many types of willing people who supply this information to the Naxals—above-ground Naxal sympathizers, servants in rich households, employees in factories, sex workers in mandal headquarters, and even mistresses of rich landlords are indispensable partners in the Naxal network.

Although financial information about others is relatively easy to obtain by local area residents, the Naxals have an edge over their rivals in one major respect. They are able to collate and tap into private information over a much larger geographical area because they have widespread networks. In contrast, locally based Naxal mimics appear severely restricted by their lack of anonymity in their local areas even if they are able to collect information about victims in that area. But the Naxals are able to accomplish two things simultaneously—anonymity in non-local areas to avoid detection as well as collection of information that aids signalling.

The Naxals are likely to have another major advantage over their rivals in collection of information. Given the economies of scale involved, and the resources that they spend in gathering information about possible victims, they may have a higher quantity and a higher quality of information over their rivals. What is more, they possibly have relatively more up-to-date information than their rivals, particularly their ex-Naxal rivals (several indeterminate

Ss and Ns in group interviews in Nallamalla).[14] Ordinary criminals simply do not have the resources to gather information on the scale that Naxals have. Even if the ex-Naxals can match Naxals in most respects, they fail when it comes to gathering the latest information about victims. Their prior experience as Naxals does not help because they lose their information networks when they exit the Naxals. Ex-Naxals take away important skills for extortion, but not the access to updated information that can be useful in passing as Naxals in an indubitable manner (indeterminate Ss, Ns in group interviews in Nallamalla).

It is increasingly the case that when the Naxals set a demand for extortion, they do their homework about the victim beforehand. This helps them signal their identities accurately. They do not set it so high that the victims cannot pay; but they set it low enough that the victims are able to pay; and yet at the same time it is enough to be worth their while to take the risk of being caught by the police and face higher charges than others.

Conclusion

This research empirically explores the role of signalling in extortions by the Naxals and the Naxal mimics in India. It discovers that while costly signalling is employed by the Naxals, deceptive mimics who do not belong to the Naxals attempt to pass themselves off as the Naxals, thereby reducing the credibility of the threats that the real Naxals issue.

There are two types of situations in which Naxal mimics succeed. First, even though they sometimes use low-cost signals and make small demands, risk-averse victims who are not able to see that it is not worth the time of real Naxals to make such small demands given their higher expected costs (punishments if caught), fall prey. Second, some mimics are ex-Naxals, who are more or less as capable of using costly to mimic signals as the Naxals, are able to successfully pass off as the Naxals.

However, the Naxals have one costly to mimic signal that no mimic, including the ex-Naxals have. They are able to signal that they have the organizational and network resources that allow them to set their demands within a reasonable range of what the victims can afford to pay, and which meets their own basic expected costs of engaging in extortion. This is a signal that consists of two components—a precise extortion price and a substantiation of

---

[14] It was hard to pin down the specific identities of subjects in group interviews.

the price with information and evidence about the victim that the real Naxals reveal to the victim while making the threat.

Because this research is the outcome of my solitary effort at collection of qualitative data from three Naxal districts in Telangana region of Andhra Pradesh there is likely to be sampling bias. My data does not cover other areas in Andhra Pradesh where, for instance, extortions involve takings of land from rich landlords and redistributing it to the poor. We learn that mimicry occurs in these types of extortions as well—where farmers who want to benefit from these redistributive extortions hire mimics from elsewhere to threaten the landlords to redistribute their land holdings. Also, my data is restricted to one-off victims. It does not extend cleanly to repeat victims.

There is potential to extend and strengthen this work using experimental means. For instance, one can think of studying the differential impact of news items about Naxal mimics and real Naxals on civilians and measure how they react in terms of willingness to bet on signals presented to them. My qualitative results predict that news reports of Naxal mimics will reduce willingness to comply with any demand while stories of real Naxal extortion will increase willingness to comply.

References

Bacharach, Michael and Diego Gambetta. 2003. 'Trust in Signs'. In *Trust in Society*, edited by Karen S. Cook, 148–184. New York: Russell Sage Foundation.

D'Souza, Shanthie Mariet and Bibhu Prasad Routray. 2010. 'Red Terror: Just How Many Crores Do the Naxals Manage to Extort?'. *Business Standard*, 29 May.

Ellsberg, Daniel. 1968. *The Theory and Practice of Blackmail*. Santa Monica, CA: RAND Corporation.

Elster, Jon. 2004. 'Kidnappings in Civil Wars'. Paper presented at the conference 'Techniques of Violence', Peace Research Institute in Oslo, 20–21 August.

Gambetta, Diego. 1993. *The Sicilian Mafia: The Business of Private Protection*. Cambridge, MA: Harvard University Press.

Gambetta, Diego. 2005. 'Deceptive Mimicry in Humans'. In *Perspectives on Imitation: From Neuroscience to Social Science*. Volume 2: *Imitation, Human Development and Culture*, edited by Susan L. Hurley and Nick Chater, 221–242. Cambridge, MA: MIT Press.

Habyarimana, James, Macartan Humphreys, Daniel N. Posner, and Jeremy M. Weinstein. 2009. *Coethnicity: Diversity and the Dilemmas of Collective Action*. New York: Russell Sage Foundation.

Mamidi, Pavan. 2010. 'Signaling and Screening for Recruitment among the Maoists in India'. *SSRN* 1688395 (https://ssrn.com/abstract=1688395)

Schelling, Thomas C. 1960. *The Strategy of Conflict*. Repr. edn. Cambridge, MA: Harvard University Press.

Shavell, Steven. 1993. 'Economic Analysis of Threats and Their Illegality: Blackmail, Extortion, and Robbery'. *University of Pennsylvania Law Review* 141: 1877–1903.

Smith, Alastair and Federico Varese. 2001. 'Payment, Protection and Punishment the Role of Information and Reputation in the Mafia'. *Rationality and Society* 13 (3): 349–393.

8

Mimicry and Its Double in the Iraqi Civil War

Stephen Holmes

'La guerre c'est la tromperie'

Ibn Khaldoun (1863: 134)

That 'Baghdad sank into sectarian violence in 2006' is well known, 'with Shia death squads storming into Sunni neighborhoods to kill civilians and Sunni suicide bombers retaliating with car bomb attacks on crowded Shia street markets' (Steele 2008, 193). Two terrifying years of sustained bloodletting followed. Then, in 2008, inter-communal conflict gradually subsided. The lull between the storms lasted for half a decade without, however, ceasing to roil and torment daily life. Reported civilian casualty rates in 2009 and 2010 were about one-sixth what they had been in 2006–2008. This steep decline in violence from the earlier period 'when some 60 to 100 bodies were being found beside the roads every morning, the victims of Sunni–Shia sectarian slaughter' (Cockburn 2008) led some commentators to speak prematurely of the end of the Iraqi civil war. But after the American withdrawal in December 2011 and eight years of one-sided Shiite ascendancy within Iraqi state institutions under Nuri al-Maliki, the flimsy truce collapsed spectacularly in 2013 (Khedery 2014). Shocking sectarian violence, in a different guise, once again engulfed the country.[1] During the summer of 2014, when the Sunni jihadists of ISIS seized control of Mosul, the monthly civilian death toll soared back into the thousands.[2] By the spring of 2015, the

---

[1] 'The levels of violence are now comparable to the dark days of 2008, though back then the death toll was falling rather than rising. They are not as bad as 2006 and 2007 when the country came close to civil war'(Harding 2014). The most recent wave of savage conflict erupted only when Iraq's battered Sunni tribes—taking advantage of the Syrian conflict next door, the power vacuum produced by the American withdrawal, and a renewed Saudi policy of funding proxies to combat the influence of Iran in the Arab world—managed to regroup and joined forces with radical Islamic militants in a shocking irredentist surge (Kirkpatrick 2014; al-Salhy and Arangojune 2014).

[2] In the month of June 2014, 'Iraq's violent insurgency has claimed some 2,400 lives—more than half of them civilians—a "staggering number,"declared the top United Nations envoy there'(UN News Centre,

Stephen Holmes, *Mimicry and Its Double in the Iraqi Civil War*. In: *Fight, Flight, Mimic*. Edited by: Diego Gambetta and Thomas Hegghammer, Oxford University Press. © Stephen Holmes (2024). DOI: 10.1093/9780191802454.003.0008

Iraqi state had effectively ceased to exist (Gardner 2015). The extreme violence and instability lasted until late 2017, when ISIS was finally dislodged from its strongholds and driven underground by Iraqi forces and a US-led international coalition (Oakford 2018).

My aim in this chapter is not to explain the tragic course of the Iraq conflict, needless to say, but solely to mine the Iraqi example to illuminate the protean forms and far-reaching consequences of mimicry under conditions of civil war. Much of the historical material on which this chapter is based comes from journalist-recorded episodes of strategic mimicry that took place during the inter-communal violence of 2006–2008. From scholarly and governmental, as well as journalistic reports, we can infer with some confidence that an eruption of Iraqi sectarian violence between the February 2006 Samarra shrine bombing and the March 2008 'Charge of the Knights' attack by Prime Minister Nuri al-Maliki's Shiite-led government on the Shiite militias of Basra was accompanied by elevated rates of strategic mimicry, defensive as well as aggressive. In 2006–2008, a veritable avalanche of newspaper articles testified that 'imposters are rife throughout Iraq' (Wong and von Zielbauer 2006). After 2008, these write-ups gradually tapered off. Not all of the earlier accounts of impersonation may have been trustworthy, to be sure. Nevertheless, the copiously documented mimicry spike of 2006–2008 provides the take-off point for the analysis that follows. I begin by focusing on the way even brief episodes of defensive and aggressive mimicry can permanently change the strategic calculations and behaviour of both models and dupes. I then turn to one of the most persuasive arguments in favour of this thesis, namely the long-standing definition of 'perfidy' as a serious war crime under international law. I conclude with a brief look at the causes and consequences of mimicry's double, namely the tactic of policemen who pretend to be members of sectarian death-squads pretending to be policemen in order to shift responsibility for their own heinous crimes.

From Camouflage to Identity Mimicry

Hiding runs a wide gamut. The simplest form of concealment is grimly illustrated by the Shia man who, while an ISIS death squad slaughtered his family, 'hid for eight hours in the middle of a stack of straw from Iraq's early summer

---

1 July 2014, accessible at http://www.un.org/apps/news/story.asp?NewsID=48177#.U7k35_mSwud); 'The blitz across Iraq has pushed the death toll there to levels unseen since the worst sectarian bloodletting in 2006 during the U.S. occupation'.

harvest, not daring to look out' (Spencer 2014). One level up, we encounter anti-predatory and predatory camouflage whereby both hunted and hunters modify their personal appearance to blend invisibly into the background hoping to avoid detection by predators and prey. In 2006–2008 Iraq, vigilantes, insurgents, and paramilitary groups on the one hand, and potential targets of ethnic cleansing and sectarian revenge on the other, shared an incentive to travel incognito. In a typical first-hand account, an Iraqi doctor, fearful of being kidnapped, reported that, when walking to the Ministry of Health in 2005, 'he wore dirty clothes and kept his hair uncombed, to avoid attracting attention' (Rosen 2009a). In 2004, similarly, female students at Baghdad University explained their decision to adopt the headscarf in a similar way: 'The whole point of wearing the scarves now was to be anonymous and unimportant, to avoid being singled out and followed, or kidnapped, or shot. It was more than a matter of blending in. It was a matter of disappearing into the landscape' (Spinner 2004).

Unable to resemble inoffensive pedestrians, professional guards wearing Kevlar and bristling with firepower extemporized their own form of camouflage. Hired to protect politicians and diplomats from assassination, foreign security contractors not only varied travel routes and schedules daily, they also changed cars, 'switching from "high-profile" vehicles (like F-350s) to "low-profile" vehicles (old sedans)' (Rosen 2007) This vehicular downgrade helped them resemble inoffensive motorists and thus avoid coming into the crosshairs of hostile snipers.

Potential targets naturally do their best to resemble non-targets. But endeavouring to slip unnoticed into the interminable traffic or a milling crowd needs to be distinguished from pretending to belong to a police or military unit or to a well-defined sectarian group not one's own. For ordinary unarmed Iraqis, falsifying one's sectarian identity was an equally natural defensive response to savage sectarian cleansing and revenge killings. Here is one account of the fatal effect of too easily legible sectarian identity signs in an area south of Baghdad known as the Triangle of Death: 'There, some of the most brutal Sunni extremists made their stand. These were monsters who threw up fake checkpoints to catch Shiites, judging them culpable because of a tape in the car or the name on an identification card. They tortured them, beheaded and mutilated their bodies' (Rubin 2009). Such blood-curdling prospects created a strong incentive for Shiites to keep forged papers and tape recordings typical for Sunnis in their glove compartments.

The chances of successful mimicry naturally decreased when debutant mimics began to be cross-examined sceptically at a traffic stop about their fake documents or trompe l'oreille music. Deceptive signalling, also during civil wars, is likely to succeed or fail depending on the accuracy of the lie-detectors in place. How foolproof these systems turn out to be hinges not only on the motivations of those assigned to discriminate mimics from models, of course, but also on their capacities which are always limited to some extent.

Trying to specify the conditions that make detection avoidance easier or more difficult helps illustrate a subtle dynamic that is well worth bringing into focus before we plunge further into the Iraqi story. Potential human prey wishing to escape prowling human predators may at first seek anonymity, as mentioned, trying to keep their heads down or 'playing dead' to avoid attracting attention. Once face-to-face inquiry is initiated, however, they will be compelled to graduate from rudimentary camouflage (seeking to avoid notice) to a more demanding kind of mimicry (pretending falsely to be a member of a specific non-prey group) under the pressure of ramped-up detection efforts. Staying guardedly silent in response to a direct question— such as 'Are you a Shi'i or a Sunni?'—would be more than futile; it would be a dead giveaway. Formulated more abstractly, impostors must devise new and more sophisticated strategies of deception when efforts to unmask impostors are seriously ramped up.

The way in which persistent grilling can compel escalation from merely withholding information to conveying a deliberate untruth was already suggestively discussed by Francis Bacon in his 1625 essay on 'the hiding and veiling of a man's self', dealing with some of the common payoffs and pitfalls of deceptive signalling. Bacon distinguished between two kinds of passing, namely, 'dissimulation, in the negative; when a man lets fall signs and arguments, that he is not, that he is' and 'simulation, in the affirmative; when a man industriously and expressly feigns and pretends to be, that he is not' (Bacon 1985, 77). Two examples of pretending to be what you are not would be a Sunni posing as a Shiite to avoid the wrath of Shiite death squads and members of a death squad pretending to be policemen to persuade credulous victims to open their front doors. A nice example of pretending not to be what you are is T. E. Lawrence wearing indigenous attire to avoid being singled out among ordinary fighters as a high-value target by Turkish snipers when leading the 1916–1918 Arab revolt against the Turks.

True, Bacon's distinction between simulation and dissimulation may seem overly refined, since you cannot pretend to be other than you are without pretending not to be what you are and vice versa. But the emphasis is different in the two cases and closely tracks the distinction between identity mimics who try to pass for specific models and mere camoufleurs who seek to melt invisibly into the background. Camouflage works by allowing the potential target to fly beneath the radar and escape the dupe's attention. Identity mimicry, by contrast, assumes that the mimic has come to the dupe's attention and therefore has to succeed in feigning a specific identity or membership in a social group not his or her own. Both camouflage and identity mimicry, as Bacon's account suggests, far from being restricted to situations of violent conflict, are ubiquitous in social life. This is true, to cite Bacon again, because both strategies not only 'lay asleep opposition' but also abet 'surprise', increasing the possibility of success and survival in hostile environments. Feigning and concealment are widely practised, in fact, because 'where a man's intentions are published, it is an alarum, to call up all that are against them' (Bacon 1985, 78).

The contest between hunters and hunted is a case in point. The human species, alongside others, evolved to pounce unexpectedly and to cope with predators that pounce unexpectedly. Two strategic purposes of deceptive signalling are to facilitate and to confound surprise attacks or, formulated differently, to let sleeping (guard and attack) dogs lie. But appearing unremarkable to elude predators or catch prey off guard will not suffice under the glare of interrogation lights. Developed to neutralize wily arts of deception, cross-examination can quickly dislodge the would-be secret-keeper from his incognito posture and propel him into a tortuous labyrinth of mendacity. This, Bacon continues, is because persistent inquisitors 'will so beset a man' who simply wants to hide his identity or intentions 'with questions, and draw him on, and pick it out of him, that, without an absurd silence, he must show an inclination one way; or if he do not, they will gather as much by his silence, as by his speech'. As a result, 'no man can be secret, except he give himself a little scope of dissimulation; which is, as it were, but the skirts or train of secrecy' (Bacon 1985, 77–78). To reformulate Bacon's distinction in the vocabulary being used in this volume: aggressors and defenders may resort to mimicry (pretending to be someone they are not) when the much simpler tactic of blending invisibly into the environment (camouflage) proves inadequate due to the out-of-the-ordinary inquisitiveness of the now wide-awake former dupes targeted by deceptive signalling.

Props

This brings us back to impersonation in civil war Iraq. The first task of a false-identity signaller, before acquiring fabricated documentation, is to conceal or dispose of genuine but incriminating documentation. Here is a recent report about how a Yazidi Kurd who 'spent years working for the U.S. Army in his area' prepared his escape from the ISIS onslaught:

> Karim had time to do just one thing: burn all the documents that connected him to America—photos of him posing with Army officers, a CD from the medical charity—in case he was stopped on the road by militants or his house was searched. He watched the record of his experience during the period of the Americans in Iraq turn to ash, and felt nothing except the urge to get to safety' (Packer 2014).

Scrapping real papers is, going back to Bacon, 'dissimulation, in the negative; when a man lets fall signs and arguments, that he is not, that he is' while presenting false papers to an inspector is 'simulation, in the affirmative; when a man industriously and expressly feigns and pretends to be, that he is not' (Bacon 1985, 77).

During the 2006–2008 phase of the Iraqi civil war, Iraqi targets of sectarian kidnapping and murder had a very strong incentive to try their luck at simulation in the affirmative: 'Ali Abdel Hussein al-Asadi, 41, an employee with Iraq's Commission on Public Integrity, said his father, a Shiite, was kidnapped from his Sadiyah home in July by men who claimed to be from the Islamic Army, a Sunni insurgent group'. So how did this Shiite abducted by Sunnis manage to survive? 'Asadi said his father had to tell his kidnappers he was a Sunni to avoid execution' (Partlow and Tyson 2007). And just as Shiites interrogated in Sunni areas extemporaneously claimed to be Sunnis, so Sunnis finding themselves endangered in Shia areas ad-libbed Shia identities. For members of both sects, revealed membership in whatever group being hounded and hunted at the moment could invite on-the-spot execution. Whether homemade or purchased, false papers offered a possible reprieve. Here is how a Shiite factory worker reported on his successful escape from the clutches of a Sunni death squad:

> One man who was released told the A.P. that the kidnappers had sorted the hostages by ethnicity, and that he had been let go because he had forged papers saying he was a Sunni. 'One of the gunmen told us to stand in one line, and then asked the Sunnis to get out of the line', said the man, who is a Shiite. 'That's what I did. They asked me to prove that I am a Sunni, so I showed the forged ID [with an identifiably Sunni name], and three others did the same. They released us'. (Burns and O'Neil 2006).

In fact, the 2006–2008 Iraqi conflict spawned a thriving market for forged identity cards, distinguishing it from the Rwandan civil war described by

David Laitin (in this volume). That provides at least some indirect evidence that strategic mimicry was widely practised in Iraq.

The affectation of normality required to hand over false papers to ruthless killers without awakening suspicions is remarkable in itself. To help maintain one's sangfroid and thereby increase the credibility of forged documents, amateur drama lessons were apparently advisable. As sectarian violence was just beginning to rage in the still residentially integrated neighbourhoods of Baghdad, friendly Sunnis and Shiites gave each other what could be called mimicry instruction: 'Relatives and neighbours meet to share knowledge on the 12 imams revered by Shias Muslims, test each other on the dates of Shia festivals or advise on where best to buy fake IDs with Shia names, in case they are challenged at gunpoint' (Ali and Poole 2006). Such reciprocal cooperation between models and mimics with the aim of duping groups of predators despised by both is genuinely remarkable and presumably quite rare.

Another report at the time confirmed the kinds of precautions Sunnis in Baghdad were absorbing to avoid being murdered on the road by Shiite zealots simply because of the names printed on their documents and other sectarian markers:

> It was fear of such a fate that led Omar, a van driver who takes food produce to markets around the city, to decide last month only to travel if he had two ID cards on him. Omar is one of the most recognisable of Sunni names, so for $25 (£13) he paid a friend with contacts in the printing industry to have another printed for him with his name as Haider, a typical Shia name. He also carries in his car a round piece of clay, which Shia Muslims place on their foreheads when they pray. A green cloth, the traditional symbol of the Shia, is kept in the glove compartment for him to place over his car's gear stick when he enters Shia neighbourhoods. 'I am fortunate because my cousin's wife is Shia so she helps me and my family learn how to act like a Shia', he said. 'I make all of us learn what she says. My children can now name all the imams and the year they were born and died. My oldest son even has a Shia religious ringtone on his mobile phone he switches to when he has to go to Shia areas'. (Ali and Poole 2006).

Arresting here, besides the futuristic archaism of religious ringtones, is the Shiism-in-three-simple-lessons training provided by 'my cousin's wife' or, more generally, by close relatives and friends. Notice also the role played by life-saving pseudonyms displayed alongside a variety of physical stage props, adopted and kept within easy reach under the tutoring of some family member or friend possessing insider familiarity with Shiite practices. Such helpful counselling suggests the importance of a legacy of exogamy and cross-denominational companionship from the Saddam era in preparing for successful defensive mimicry during the initial phases of savage inter-communal violence. Sunnis counted on Shiite friends and relatives for crash courses in detection-resistant Shiite identity signals, and vice versa.

In 2006–2008, in any case, forged documents were only one item in the impostures' mimicry kit. Sunnis wishing to pass for Shiites were able to consult a website listing a variety of how-to-mimic and where-to-shop-for-props suggestions:

1. Practise imitating another personality and have an ID with another name (you can get these forged IDs from Muraidy market in Sadr city), especially if your name was Omar or Othman and if your family name was Dulaimy or Janabi, or if your birthplace was in one of the Sunni-majority cities.
2. Memorize the names of the twelve imams.
3. Learn to pray in the Shia way and carry turba [Shia holy clay] in your pocket.
4. Keep a turba in your house where it can be seen, and put up if necessary a black or a green banner on the roof.
5. Keep a poster in your house of Imam Hussein. You can buy them in Mutanabi Street in Baghdad.
6. Keep a copy of the *Sajadi* newspaper [a Shia paper that has Shia prayers] and read some of the prayers, some of them are touchingly beautiful.
7. Keep a latmiya [Shia song] on your mobile phone.
8. Learn how to curse Yazid and Maawia and Bani Omaya [early Sunni caliphs hated by the Shias] and in the way the Shias do.
9. Wear or keep black clothes in your house, especially in ceremonies that demand it.
10. Learn about the different Shia ceremonies (the death of the imams, their birth, and the joy of Zahraa).
11. Pray in a husseiniya or a Shia mosque. Remember that Shia and Sunnis are not enemies, but there are misled, ignorant people and victims of evil plans who want to spread the breath of hostility in Iraq.(Hider and al-Hamdani 2006).

We can retro-engineer this list to inventory the identity markers, visible signs customarily associated with unobservable group membership, used by Shiite death squads to cull Sunnis from the crowd. The inability to name all twelve imams, for example, could be a death sentence for captured Sunnis. The urgency of obtaining forged papers to replace authentic ones was reinforced by daily news bulletins: 'One morning 14 bodies were found, all with ID cards in their front pockets identifying them as "Omar", a Sunni name ... And Sunni militias were retaliating, stopping buses and demanding the *jinsiya*, or ID cards, of passengers and executing those with Shia names' (Rosen 2006). As a consequence, both sides associated credible identity signals, genuine or bogus, with the difference between skin-of-one's-teeth survival and sudden death.

Defensive Mimicry

In the fog of war, combatants commonly mistake hostiles for friendlies and friendlies for hostiles. Such a naturally bewildering environment does not presuppose deceptive signalling but it makes deceptive signalling more likely to succeed. This is true of civil wars as well, even though combatants in such conflicts presumably know more about each other than do soldiers in international wars. When the civil war broke out in earnest in 2006, Iraqis who had lived for decades in relatively insulated face-to-face communities were suddenly tossed into a vortex of mass displacement. Ousted from their traditional localisms and forced to move across unfamiliar terrain, many ordinary Iraqis were suddenly compelled to interact on a daily basis with total strangers. This exceptional environment made widespread identity mimicry practicable for the first time. Reducing the value of life-long acquaintance as a basis for interpersonal trust, the displacement crisis also coincided with the emergence of both Sunni and Shia death squads engaged in tit-for-tat sectarian atrocities. Just as residential displacement made identity-mimicry feasible, the likelihood of being murdered made identity-mimicry worth trying.

How levels of strategic mimicry vary inversely with levels of interpersonal familiarity is nicely illustrated by a temporary exception to the general trend towards residential segregation in the initial phase of Iraq's civil war. In Baghdad, the neighbourhood of Bab al-Sheik was 'spared the sectarian killing that has gutted other neighbourhoods, and Sunnis, Shiites, Kurds and Christians live together here with unusual ease'. A multi-sectarian holdout, the area was so safe, even at the height of inter-communal violence in 2007, that American journalists could saunter down the streets without guards. Violence was minimal in the neighbourhood 'largely because of its ancient, shared past, bound by trust and generations of intermarriage'. One inhabitant, named Waleed, explained how intimacy among insiders, who knew everything there was to know about each others' families, provided an effective shield against violent intrusions by outsiders. Waleed's son was killed when he moved temporarily to another neighbourhood where his 'powers of discernment' were diminished by lack of face-to-face familiarity: "'We didn't know each other's backgrounds", said Waleed, sitting with Monther in a barbershop in Bab al-Sheik, rain spitting on the street outside. Neither man wanted to be identified by their last names out of concern for their safety. 'Here, he can't lie to me', he said, jabbing a finger in Monther's direction. 'He can't say, "I'm this, I'm that", because I know it's not true' (Tavernise and Hilmi 2007).

You could falsely claim that 'I'm this, I'm that' only to interlocutors who were never your neighbours and who are not personally acquainted with your family background. Hence, widespread resort to mimicry is unlikely to occur in stable, tribal, kinship-based societies, unless a violent uprooting has recently taken place. This was precisely the case in Iraq.[3]

Explaining the decline of reported incidents of defensive mimicry after 2008 can help us identify the conditions that made mimicry possible and desirable in 2006–2008. One reason for the eventual fall-off in reported cases of mimicry might possibly be *learning* by the previously cozened population. After repeated exposure to strategic deception, infuriated dupes may have improved their capacity to spot forged documents and see through clever disguises. Reducing the gullibility of the intended targets of deceptive signalling, would also reduce the incentive for endangered parties to assay identity mimicry, whatever level of danger they face. This sort of explanation is what James Fearon (in this volume) has in mind when arguing that 'if detection efforts are moderately accurate, then the equilibrium amount of mimicry may be quite low even in a high stakes case like genocide'.

In the Iraqi case, however, the gradual stabilization of residential segregation probably played a more important role than improved detection efforts in reducing the possibility and desirability of identity mimicry. As Nir Rosen (2009c) reported at the time, by 2008 there were 'fewer people dying because there were fewer to kill; the cleansing had nearly been completed, with Sunnis and Shias separated in walled enclaves run by warlords who had consolidated control'. As members of each sect began to retreat for protection into relatively homogeneous sectarian enclaves, interpersonal familiarity began to reassert itself, reducing both the need and the opportunity for Sunnis and Shiites to pretend to be Shiites and Sunnis.

Ironically, vestigial Saddam-era friendships between Sunnis and Shiites also accelerated the move towards residential segregation as some Sunni and

---

[3] Notice also that the informants quoted above were afraid to have their names printed, a reticence suggesting the continued importance of anonymity as a protective shield when your neighbourhood, however peaceful internally, is being circled by ravenous wolves.

Shiite friends voluntarily traded houses with each other so that each of their families could live in safer, all-Sunni or all-Shiite areas. While such residential segregation reduced the incentive to falsify one's sectarian identity, it also reduced the day-to-day contact between groups that had earlier facilitated the protection of friends and family members across sectarian lines. And it presumably made it much harder for members of hunted groups to learn from their friends and family members across the sectarian line how to credibly feign identities not their own.

It is nevertheless correct to say that successful mimicry of sectarian rivals by members of a hostile denomination initially depends upon and ultimately erodes naïve credulity. But false-identity signalling not only causes an intensification of uncovering efforts, replacing naiveté with wariness. It can also inject a paranoid jumpiness into the machinery of detection. It can turn sensible vigilance into pathological suspiciousness. It can make identity auditors doubt the authenticity of genuine identity signals, wrongfully suspecting models of being their own mimics in disguise.

Hamudi Naji, a mafia-like Mahdi Army commander responsible for murderous attacks on Sunnis, made precisely this mistake. He was furious with a certain Captain Mushtaq, an Iraqi Army officer, for having obstructed Naji's sectarian murder campaign. Therefore:

> Naji arranged for Mahdi Army men and some Iraqi national police to go to Mushtaq's house, but they went to his neighbor's house by mistake. The posse insisted that the neighbor's ID card was fake, and they put him in the trunk of their car. Naji called Mushtaq's phone and was surprised when Mushtaq answered. 'Who are you?' he asked. 'Mushtaq', the captain replied. 'So who is the lamb we have here?' Naji asked, using slang for a victim about to be killed. The neighbor was released, terribly beaten. Mushtaq sent some of his family to the south and his wife and children to Egypt (Rosen 2009c).

Having been fooled once too often by forged IDs, the hyper-suspicious Naji insisted erroneously 'that the neighbor's ID card was fake'. So how are innocent models likely to react when abused and threatened in this way by paranoid former dupes infuriated by the deceptive practices of the models' mimics?

This is an important question, downplayed by analyses focused solely on the duet between mimic and dupe. Arguably, the key to mimicry's significance as a social practice resides in the non-trivial harm that mimics intentionally or unintentionally cause to be inflicted on their models and the harm-mitigation strategies adopted by the unjustly burdened models in turn. A possible example is the CIA's decision to disguise its information-gathering mission in Abbottabad as an effort to vaccinate children. When the mimicry was discovered, 'angry villagers, especially in the lawless tribal areas on the Afghan border, chased off legitimate vaccinators, accusing them of being spies' (McNeil 2012).[4]

In the Iraqi civil war, a brief period of defensive mimicry, 2006–2008, gave both Sunni and Shiite models a strong motivation to manufacture identity signals that would be harder to imitate in the future. Models harried by hyper-suspicious attempts at unmasking impostors have a good reason not only to cluster in homogeneous enclaves, but also to develop new identity signals that are more difficult to imitate. That is why identity mimicry can be causally important even when, after flaring up briefly, it falls back to low levels in response to moderately improved rates of detection. Temporary strategies adopted by models to devise theft-proof identity markers when mimicry rates are temporarily high can crystallize into enduring habits of the heart.

Defensive mimicry by Sunnis was bound to make Shiite death squads suspect even genuine Shiites of being Sunnis in disguise. Afflicted by such externalities of strategic mimicry, the model will not sit idly. Even as both Sunnis and Shiites retreated into residentially segregated areas, members of each group needed to let their own sect's militiamen know that they were genuine coreligionists rather than impostors. They needed to communicate their unobservable membership status in a more eye-catching and convincing way than was necessary in the age of peaceful inter-communal coexistence. This is arguably how an interim uptick in defensive mimicry, which contracted to low levels 'in equilibrium', nevertheless left behind a substantial hardening of sectarian identity in Iraq. This is of no small consequence since a radical intensification of sub-allegiances fostered the fateful dismantling of non-denominational Iraqi identity that, in 2015–2017, led to the collapse of Iraq as a unified state.

Aggressive Mimicry

Levels of defensive mimicry, on which we have been focused so far, are highly responsive to the accuracy of detection efforts. This can also be true of some forms of aggressive mimicry, especially when it involves 'sleeper cells' who must live undetected alongside their targets for a relatively extended period

---

[4] See also *Scientific American* (2013).

before striking.[5] But it is much less true of the most common form of aggressive mimicry in wartime, where a predator gets a momentary jump on its prey by resembling for a few seconds either a non-threatening party or a member of the prey's own group. In the run up to waylaying its prey, an aggressive mimic may have to deceive the unsuspecting victim only for an operationally decisive blink-of-an-eye. If the target can be fooled for the instant it takes to open his front door or let down his guard, the fatal shots will have been fired. Wearing a lookalike uniform to gain a split-second advantage in a terrorist ambush is very different from sustaining a false identity for months at a time to garner useful intelligence, Donnie-Brasco style, inside a hostile organization.

Posted sentinels who occasionally slack off from boredom or exhaustion can be savagely punished when caught, as they were for example in the Roman legions, but periods of relaxation and distraction will inevitably occur, and that is what gives the aggressive mimics their opportunity, even against a potential dupe who has the greatest possible incentive to avoid being deceived. According to Fearon (in this volume): 'The equilibrium rate of mimicry is determined not by the mimic's preferences, but by how much the targets care about catching them'. But the *capacity* of the targeted dupes is just as important as their *motivation*. However much 'they care', sentinels whose reflexes are inevitably dulled by routine cannot be expected, 100 per cent of the time, to distinguish mimics from models faster than mimics can take advantage of a momentary ruse.[6] In Iraq, moreover, as in Afghanistan, some aggressive mimics will not be greatly deterred by the potential costs of failing to pass. This is true when being killed by alert sentinels qualifies the attacker as a glorious martyr in any case.

---

[5] Clandestine ISIS agents slipping into Kurdish-controlled territory, posing as refugees from the ISIS onslaught, and plotting a surprise attack could presumably be detected by Kurdish identity-monitors and future missions thereby deterred: 'Some Kurds have become deeply suspicious of their Arab neighbours overnight. The region has given refuge to tens of thousands of Arabs displaced by violence in the rest of the country since the start of the year. "We opened the doors to the Arabs but yesterday we discovered that in some houses they have weapons", said Ziad Taha Aziz, 44, who sells shoe polish and brushes at a stall in the market. "Some of them are good, and some of them are bad. We need to arrest them all and see whether there are traitors among them. We think there are sleeper cells"' (Reuters 2014).

[6] Cf. the infamous 30 December 2009 suicide attack at a forward operating base, Camp Chapman, near Khost, Afghanistan. Seven American agents responsible for picking drone targets inside Pakistan were killed along with a senior Jordanian intelligence officer. The suicide bomber, in a pre-recorded video, claimed to be undertaking the attack in retaliation for the targeted killing campaign being overseen at the base. He had apparently spent months lowering the guard of the US agents by providing intelligence on the whereabouts of low-level operatives who were duly killed by drones. The sloppy screening that, on the day of the attack, allowed the infiltrator to come lethally close to so many top agents was allegedly a gesture of courtesy stimulated by his promise to deliver actionable intelligence on the whereabouts of al Qaeda's number two, Ayman al-Zawahiri. Although the agency's most skilled drone targeters had a powerful incentive to distinguish accurately between friend and foe, not even their well-honed deception detectors could function infallibly 100 per cent of the time.

Uniforms

Uniforms simulate uniformity. They hide real differences behind a veneer of sameness. In the age of mass armies, they make it easier for soldiers to distinguish hostile from friendly forces. Reliance on such a childishly simple signal of otherwise unobservable group membership, however, creates an irresistible opportunity for mimics. Indeed, treacherous killing by combatants wearing deceptive clothing remains the most consequential form of wartime mimicry.

Sometimes combatants gain an advantage by imitating non-combatants, as when US Special Forces in Vietnam wore indigenous conical straw hats and dressed as fisherman while travelling in sampans, or when 'Hamas militants are fighting in civilian clothes; even the [Gaza] police have been ordered to take off their uniforms' (Erlanger 2009), or when 'Three Israelis posing as tourists in Bethlehem opened fire today on a group of Arabs who threw rocks at a nearby army jeep, killing one of the Arabs' (*New York Times* 1989), or when Sayeret Matkal units, including Ehud Barak, entered Lebanon 'dressed as women' to kill three senior Palestine Liberation Organization's (PLO) members in retaliation for the Munich massacre.[7] According to Lt. Col. Erez Wiener, Israel Defense Forces (IDF) division operations chief for the West Bank region: 'Of the 2200 arrests Israelis made in the West Bank last year, about 1500 were conducted by special operation forces (SOF), including 366 by a special Arab impersonation unit', he explained. 'These specially selected and trained personnel have demonstrated the ability to completely blend into the opposing population for intelligence and operational purposes'.[8]

The deceptive wearing of uniforms to catch the enemy off guard was also quite common in 2006–2008 Iraq. Here is a characteristic report:

Gunmen dressed in Iraqi Army uniforms kidnapped an Iraqi banker and his son from their house in Baghdad, after shooting five of the family's bodyguards execution-style in the backyard, the police said Friday ... As many as 14 gunmen, all in uniform and some in masks, climbed over the outer wall of the house, knocked on the back door and explained to the family that a mortar had been shot from the area and that they needed to search the house, a friend of the family said. They began to search the house and then led Mr. Kubba [the banker] away with his son. The family assumed the men were Iraqi soldiers. 'They looked so convincing', said the friend, who had visited the men's wives on Friday night. He spoke on

---

[7] Operation Aviv Neurim, conducted on 9–10 April 1973. http://www.jewishvirtuallibrary.org/jsource/History/opspring.html.

[8] http://defensetech.org/2005/03/08/israeli-impersonators-key-to-terror-fight/.

condition of anonymity because the family had not authorized him to discuss the case. Hassan Kubba's wife 'took it fully that they were friendly forces', the friend said. When the gunmen killed the five guards, lining them up against the outer wall of the house and shooting each in the head, the family realized that they were not who they said they were. Phone calls to the army and the police confirmed their fears (Tavernise 2006).

Combatants presumably began to wear cloned copies of their enemy's uniforms as a ruse to facilitate surprise as soon as uniforms, recognizable at a distance, came into standard use. The practice would never have been singled out for special prohibition by the laws of war, as we will see below, if it had not been disturbingly widespread. The deadly hoax, in any case, remains ubiquitous and commonplace today. For instance, in Iraq itself: 'Sunni Muslim militants, mostly disguised in army and police uniforms, struck at polling centers around Baghdad and northern Iraq as militants tried to disrupt Iraq's fourth national election since the fall of Saddam Hussein in 2003' (Rasheed and Raheem 2014). Similarly, reports of insurgents trying 'to disguise themselves in coalition military uniforms' are filed almost daily from Afghanistan (Sieff and Salahuddin 2014). Similarly, in Israel, 'Eight Palestinian militants emerged from a tunnel some 300 yards inside Israel on Saturday morning, armed with automatic weapons and wearing Israeli military uniforms' (Barnard and Rudoren 2014). And in Tunisia, with a rapacious rather than homicidal intent, 'Militants wearing police uniforms on 16 February set up a false checkpoint in Ouled Manaï, a village in the north of Jendouba, where they robbed and extorted passing vehicles.'[9] Examples could easily be multiplied.

Dressing in the enemy's uniform, however loudly deplored, is a customary military ploy. At key moments during the Second World War, German commandos donned Danish, Dutch, Belgian, Russian, American, and British uniforms as *ruses de guerre* to gain a momentary advantage during a surprise attack (J. S. Lucas 1985). Late in the war, American OSS agents returned the favour, dressing as German officers in the run-up to the Allied advance into Axis territory (Persico 1979). Indeed, 'State practice shows that governments have been willing to deploy Special Forces in civilian attire or enemy uniforms where a major advantage is anticipated, and where the gain is greater than the risk to the deployed personnel' (Parks 2003, 546). A good example is recorded in the famous post-Second World War trial of a German officer,

---

[9] 'Jihadist use of false checkpoint in Jendouba signals heightened security risks on Tunisia's roads,'in IHS Jane's Country Risk Daily Report (18 February 2014).

Otto Skorzeny, for leading a German Panzer Brigade into combat dress in American military uniforms. (*New York Times* 1947).

Fake uniforms were also worn by all sides in Iraq's civil war. Journalists reporting from Iraq in 2006–2008 began to stress the sartorial deceit apparently deployed by a range of Iraqi death squads. Police and army uniforms are easily falsified signs. Thus, 'kidnappers wearing Iraqi police uniforms conducted a mass abduction at Iraq's ministry of higher education in Baghdad', and 'Iraqis wearing police commando uniforms kidnapped a group of British contractors at the ministry of finance in Baghdad in late May 2007' (Katulis 2007). Military, as well as police uniforms, were reportedly used in lethal masquerades: 'at least 100 gunmen in Iraqi Army uniforms kidnapped several senior Oil Ministry officials from their homes in a fortified government compound' (Cave 2007). Similarly, 'Men described by witnesses as Sunni insurgents dressed as Iraqi Army troops stormed a Sunni village on the southeastern outskirts of Baghdad at dawn on Thursday, killing at least 11 people during a three-hour firefight before American and Iraqi soldiers drove them off … The attackers wore Iraqi Army fatigues … along with Awakening Council uniforms stolen during previous raids' (Buckley 2007).[10] Similarly:

> Gunmen wearing Iraqi army uniforms shot and killed 15 men Saturday in a Kurdish Shiite village northeast of Baghdad … The attack against the villagers occurred early Saturday when gunmen wearing army uniforms entered the village of Hamid Shifi, about 60 miles northeast of Baghdad. They rousted families from their homes and opened fire on the men, killing 15 of them (Reid 2007).

The tactical value of stolen or purchased uniforms to gain a jump on prey was confirmed by reports from eyewitnesses, such as this one from a woman watching from her window: 'From the back of an SUV… the men in army uniforms hauled out a blindfolded passenger, who appeared to be alive, and moved him to the trunk of the sedan. Then the men shed their army uniforms, tossed them into the vehicles and drove away' (Filkins 2008, 322).

Alongside their bogus uniforms, the killers often adopted tricked-out modes of transportation: 'Victims of Baghdad's violence are often taken away by men in police uniforms, and sometimes in police vehicles, and later found dead' (Knickmeyer 2006). The gangland-style murder of a Sunni leader was described as follows: 'at least 10 vehicles that appeared to belong to the Iraqi army stopped outside the western Baghdad house of Kadhim Sarheed Ali al-Dulami, a sheik of the Sunni al-Dulami tribe, before gunmen went inside

---

[10] The Awakening Council refers to the Sunni tribal grouping allied with US forces.

the home and shot the men' (CNN 2005). Stories in the press repeatedly claimed that sectarian death squads, posing as the Iraqi police, hijacked, borrowed, or shrewdly cloned police cars and vans.[11] Teams of killers also lured and snared their victims with improvised roadblocks: 'Sunni gunmen ambushed a convoy of minibuses Saturday night at an improvised checkpoint on the dangerous highway south of Baghdad, killing 10 Shiite passengers and kidnapping about 50' (AP 2006). Illustrating how defensive mimicry can serve, or morph into, aggressive mimicry, fake uniforms can help killers pass through genuine checkpoints undetected and thereby to gain access to otherwise barricaded facilities where they can slaughter their enemies. By 2006, 'Baghdad and the surrounding region [were] in the grip of a ferocious sectarian conflict between rival Sunni and Shiite factions. Dozens of people [were] killed every day, many of them by gangs in police uniforms and vehicles' (AFP 2006).

Seeking immunity from attack, as discussed earlier, defensive mimics can foreseeably but unintentionally redirect predatory violence onto their models. Aggressive mimics may do the same, but they may also do so intentionally as a calculated stratagem. For instance, insurgents who seek a short-term tactical benefit by using police or military uniforms and vehicles while committing atrocities may also hope to foment popular distrust towards the genuine soldiers and policemen. Public suspicion of all uniformed officials can provide a collateral benefit to insurgents alongside the tactical advantage of duping victims in the run up to an attack.

When criminals and terrorists commonly wear fake uniforms, average citizens are caught in a bind. If they stop for police impostors at a checkpoint, they may be shot; but they may also be shot if they fail to stop at a checkpoint for real policemen. This nuts-making uncertainty presents operational problems for the authorities as well. States cannot function normally when personally unfamiliar but officially outfitted police officers are suspected of being murderers in disguise as soon as they appear on the horizon.

The first solution that suggests itself, naturally, is redesigning uniforms to make them more difficult to replicate: 'Bombers in Iraq have routinely disguised themselves in uniforms to bypass security checks. The problem was so prevalent in 2006 that the US military redesigned Iraqi federal police uniforms after the old one was copied by anti-government fighters, death squads and common criminals' (Aljazeera 2010). To counter impersonators, Iraqi officials, too, made their own repeated efforts to introduce

---

[11] Does the fact (if it is a fact) that plainclothes policemen often ride around in unmarked cars make impersonation easier or pointless?

imitation-proof uniforms: 'new blue, black and gray design—in a digital camouflage pattern'—for the Iraqi national police force. 'Iraqi police modeled
new-look uniforms on Monday in a bid to stay a step ahead of the death
squads they claim are masquerading as officers to carry out sectarian and
political murders'. The Iraqi Interior Minister declared that these redesigned
outfits would be 'difficult for assassins to fake' (AFP 2006). He 'repeated
promises made since early this year that police would soon be issued uniforms and vehicles that would be difficult to duplicate' (Knickmeyer 2006).
He even stated outright that: 'These new garments will not be counterfeited'
(AFP 2006). Unfortunately, the counter-counterfeiting attempt did not prove
especially successful: 'A series of incidents in which terrorists posed as police
spurred the Iraqi government months ago to say that it would reissue police
uniforms to make them harder to copy, but it hasn't acted on the issue. Fake
uniforms can easily be purchased on the street, officials said'. (Vanden and
Sabah 2006). That genuine uniforms were also stolen and sold is not unlikely.
In any case, new-modelled uniforms will make impersonation more difficult only if careful records are kept and regularly consulted: 'Police Col.
Abdul-Munim Jassim, proudly wearing his new uniform, told AFP why it
would be difficult for assassins to fake. "The Americans take a photo of the
policeman together with the number of the uniform. If found elsewhere,
it will immediately be recognized as stolen"' (AFP 2006). Such attempts to
prevent routine duplication of official uniforms proved consistently futile,
however.

Perfidy

As American troops poured into Iraq in 2003, 'Saddam's Fedayeen and their
allies had been dressing in civilian clothes to get close to US troops, sometimes even faking surrender, only to open fire at short range' (Baker 2003).
Another report tells of 'two separate ambushes where Iraqis, some in civilian clothes, pretended to surrender, only to open fire when the Marines
approached' (Dillon 2003). Under the laws of war, such treacherous killing is
classified as the war crime of 'perfidy', an offence with which American prosecutors charged several suspects during the War on Terror (C. Rosenberg
2014). Abd al-Hadi al-Iraqi, an Iraqi prisoner who commanded insurgents
during the war in Afghanistan, was charged before a US military tribunal
with the crime of 'perfidy' and remains in Guantánamo Bay even today (C.
Rosenberg 2022). Classic examples of perfidy include transporting belligerents in ambulances to conduct surprise attacks, pretending to be wounded

to draw the enemy into shooting range, and holding up a white flag with the aim of lowering the adversary's guard. Dressing as civilians or wearing the uniform of the enemy during combat are two other examples. According to Article 39.2 of the 1977 Additional Protocol to the Geneva Conventions, in particular, the illegality of perfidy means that 'It is prohibited to make use of the flags or military emblems, insignia or uniforms of adverse Parties while engaging in attacks or to shield, favour, protect or impede military operations'.[12] An earlier attempt to discourage dressing to resemble the enemy in combat appears in the 1863 Lieber Code: 'Troops who fight in the uniform of their enemies, without any plain, striking, and uniform mark of distinction of their own, can expect no quarter'.[13] Article 23 of the Hague Convention IV (1907) declares more vaguely that it is forbidden to make 'improper use' of 'the military insignia and military uniforms of the enemy'.

In 2012, US military prosecutors swore out an eight-page charge sheet for Ali Musa Daqduq, accusing him of perfidy among other war crimes:

> Mr. Daqduq is accused of conspiring with several groups, including the Quds Force of Iran's Islamic Revolutionary Guards Corps, to train Shiite militias to use roadside bombs and other insurgent tactics. The most serious charges stem from what prosecutors say was his role in helping organize a raid in January 2007 by insurgents who wore American-style uniforms and carried forged identity cards. They killed five American soldiers in Karbala, Iraq; one in the raid, and four others who were captured and whose bodies later dumped by a road (Savage 2012).[14]

Charging Iranian-backed Iraqi insurgents with this particular war crime may seem hypocritical in light of the common claim that US Special Forces regularly fight out of uniform.[15] Crucial for our purposes is not the inconsistency of US military justice, however, but what the social science of mimicry can learn from the legal analysis of perfidy. The prohibition of perfidy in the laws of war may or may not have seriously reduced the willingness of combatants

---

[12] Protocol Additional to the Geneva Conventions of 12 August 1949, and relating to the Protection of Victims of International Armed Conflicts (Protocol I), 8 June 1977.

[13] Art. 63; see also Art. 83: 'Scouts, or single soldiers, if disguised in the dress of the country or in the uniform of the army hostile to their own, employed in obtaining information, if found within or lurking about the lines of the captor, are treated as spies, and suffer death'.

[14] Mr Daqduq's eventual release by the Maliki government was generally taken as a sign that, after the US withdrawal, Iran had greater influence over Baghdad than Washington (Gordon 2012).

[15] 'The distinction between combatants and non-combatants is also threatened by the practice of US special forces, which constitute an increasingly important part of the US military yet have—with the apparent support of Secretary of Defense Donald Rumsfeld—taken to wearing civilian clothing'(Byers 2006). Suspicions that British concern for *tu quoque* arguments has also diminished since the post-Second World War trials are confirmed by the report that, in 2005, two British 'undercover officers dressed as Iraqis'were arrested by Iraqi police and were 'then freed as a British armored vehicle blasted through the wall of their jail after an angry crowd began rioting outside'(Tavernise 2005).

to resort to treacherous killing. But it is nevertheless noteworthy because of the particular revulsion it reserves for one particular form of wartime deception. As a general matter, the laws of war make little attempt to regulate *ruses de guerre*. As Article 37.2 of Protocol I to the Geneva Conventions explicitly states:

> Ruses of war are not prohibited. Such ruses are acts which are intended to mislead an adversary or to induce him to act recklessly but which infringe no rule of international law applicable in armed conflict and which are not perfidious because they do not invite the confidence of an adversary with respect to protection under that law. The following are examples of such ruses: the use of camouflage, decoys, mock operations and misinformation.

Other permitted forms of potentially lethal deceptive signalling in wartime are false retreats to lure enemies into an ambush where they will be killed or captured, artifices that make one's forces seem larger or smaller than they are, removing or shrewdly repositioning signposts, and issuing false orders to enemy troops using intercepted enemy codes and passwords. A non-judgmental attitude towards such wartime deceits may simply be a concession to reality. Not only Ibn Khaldun, as cited in the epigraph, but almost all theorists of war and civil war including Sun-Tzu (2006, 42): 'War is founded on deception'; Machiavelli (1996, 299): 'he who overcomes the enemy with fraud is praised as much as the one who overcomes it with force';, Hobbes (1968, 188): 'Force, and Fraud, are in war the two Cardinal virtues', and many others agree that war inevitably involves deception. Ulysses already explained the reason succinctly in Sophocles' play *Philoctetes* (1953, 167), first performed in 409 BC:

NEOPTOLEMUS: Don't you believe it wrong to tell a lie, sir?
ULYSSES: No, if success and safety depend upon it.[16]

So why, given this ancient and ongoing tolerance for wartime deception, do the laws of war single out perfidy as an especially heinous breach?

One reason might conceivably be the sickening consequences for those who are duped by perfidious operations. The 1995 massacre at Srebrenica provides a terrifying example.

---

[16] To be fair, this accepting attitude towards wartime deception contrasts with a hortatory tradition, going back to antiquity, according to which gallant souls scorn killing treacherously, wishing to prevail by valour and force not by craft and cunning. For instance, 'no true man of spirit deigns to kill his man by stealth'(Euripides, *Rhesus*, line 510) (Lattimore and Grene 1958, 27).

Bosnian Serb soldiers wearing stolen UN uniforms and driving stolen UN vehicles announced over megaphones that they were UN peacekeepers and that they were prepared to oversee the Bosnian Muslims' surrender and guarantee they would not be harmed. Disoriented and exhausted, many Bosnian Muslims fell for the lie. It was only after they had surrendered that they discovered their fatal mistake. For in surrendering, they were going to their deaths. Those whom the Serbs got their hands on were killed by firing squad (Rhode n.d.).

Particular sensitivity to the fate of those duped by false uniforms is understandable in a case like this where unarmed civilians were treacherously killed by combatants disguised as peacekeepers. But what about conflict between armies where only soldiers, each side intent on killing the other, are involved? Why is perfidy distinguished so sharply from other equally fatal forms of deceptive signalling, all of them clearly designed to elicit the confidence of the enemy with the intention of gaining a life-or-death advantage from the adversary's gullibility?

International lawyers elaborating on Article 37.2 explain the especial treacherousness of perfidy, rather law-bookishly, by emphasizing that the confidence elicited and breached by the mimic has been enshrined in international humanitarian law.[17] The attacker who treacherously feigns injury or surrender is wickedly exploiting the adversary's willingness to give quarter in compliance with international law. A classical example comes from the Second World War, when some Japanese soldiers reportedly hoisted the white flag to entice Americans into the open where they would be easy targets for snipers. The predictable result was the American forces ceased giving quarter and starting killing on sight Japanese raising white flags, just as the apparent Japanese practice of booby-trapping their own wounded soldiers, allegedly led American troops to shoot and kill wounded Japanese combatants (Linderman 1999, 152; Straus 2003, 116).

For international lawyers, in any case, 'perfidy' is exceptionally wicked because it is a direct attack on international law itself. It violates the specially protected status granted by International Humanitarian Law (IHL) to certain combatants as well as most non-combatants. Social science, with no professional investment in the binding power of international law, can reformulate the point less doctrinally by emphasizing the way aggressive mimicry predictably causes the dupe to respond to the mimic's violence by attacking the mimic's model. Intentionally or unintentionally, by awakening chronic

---

[17] 'The essential element of perfidy is, accordingly, that the confidence induced (and broken) must relate to the protected status under the rules of international humanitarian law'(Moir 2009, 516).

distrust in the group whose members were taken off guard by mimicry, wartime impostors regularly expose their innocent models to lethal attack.

Defensive mimicry, too, as we have seen, unloads some injurious externalities onto the mimics' models. Interestingly, IHL does not regard wearing the uniform of the enemy or dressing as a civilian in order to escape a prisoner-of-war camp as perfidy, even though it is a form of deception that is likely to burden civilians in the vicinity. Once the escapee has rejoined his unit and reported on the enemy's position, moreover, such a successful ruse is bound to pose a lethal threat to the dupe as well. This doctrinal distinction between the lawful and the unlawful wearing of the enemy's uniform, however shaky the line, closely tracks the social scientist's contrast between defensive mimicry, which burdens the innocent models only unintentionally, and aggressive mimicry which, besides aiming to kill the dupe can also adopt the higher-order strategic objective of harming members of the dupe's community by luring them into a self-defeating attack on innocent models.

Notice, too, that IHL does not consistently classify 'playing dead' as a form of perfidy. The international lawyer would explain this by saying that dead bodies escape being targeted by the enemy (and are therefore attractive models for living soldiers to imitate when under fire on a battlefield) not because of any respect for protected status granted by IHL, but only because they are seen as posing no threat. The social scientist would instead emphasize that while such mimicry may harm the dupe, by adding a survivor to the enemy's forces, it will have no dire externalities for the model who, being dead, is beyond harm.

Perfidy is singled out among the ordinary *ruses de guerre* for particular opprobrium, from this perspective, not because bad faith is morally dishonourable or because *hosti etiam servanda fides* ('faith must be kept even to the enemy'), but because treacherous killing effectively sabotages the dupe's cognitive capacities, destroying his ability to differentiate clearly between the harmless and the dangerous. The resultant fear of false negatives naturally raise the rate of false positives:

> Once US troops realised that enemy soldiers might dress in civilian clothes and drive civilian vehicles, they had to treat all private cars and vans as potentially hostile. That does not justify their decision to shoot dead a group of unarmed women and children on Monday. But would the incident have taken place if an Iraqi officer had not lured the Americans to their death, two days earlier, by pretending to be a taxi driver? (Rozenberg 2003).

Fear of being fooled twice unleashes the dupe's pre-emptive strikes upon the mimic's guiltless models.

Most significantly for the social science of wartime mimicry is the possibility that impostors intentionally engage in such identity-faking in order to induce the dupe to inflict grave harm upon the model being imitated with the expectation that such injury to the innocent, however unintentional, will create a self-defeating backlash grievously harmful to the dupe itself. When it involves soldiers mimicking civilians, for example, perfidy makes it psychologically difficult for dupes to respect the all-important *jus in bello* principle of distinction.

By luring the dupe into killing non-combatants for fear that they are combatants, insurgents who fight in civilian clothing can, among other things, undermine the sense of moral superiority natural to a regular army fighting against barbarian 'terrorists', defined as those who kill non-combatants in pursuit of political aims. This is already some kind of perverse moral victory. Considerations of this sort help explain why, in the laws of war written by states whose soldiers wear uniforms, not by guerrilla movements whose combatants often fight under cover, 'It is prohibited to kill, injure or capture an adversary by resort to ... The feigning of civilian, non-combatant status'.[18] It is inevitable that soldiers threatened by death at the hands of enemy combatants dressed as non-combatants will be drawn into killing non-combatants whatever the laws of war demand. To say, in this context, that the dupe 'has agency' is to say that his behaviour can be readily manipulated in morally repugnant but perfectly predictable ways. The strict prohibition on deliberately targeting non-combatants may have originated as a strategy for commanding officers to direct their troops away from soft targets, such as undefended towns containing women and booty, and towards dangerous enemies with the motive and capacity to kill them. But it also has the militarily useful side effect of reducing the degree of civilian hostility to an invading army. If insurgents can prevent their enemy from respecting the combatant–non-combatant distinction, they can hold up the bodies of the mistakenly killed to rally domestic and international support for their cause.

If lifelong inhabitants of Baghdad could easily be misled once they debouched from their well-demarcated ancestral neighbourhoods, one can imagine how difficult it must have been for American ingénues—parochial, unsophisticated, and monolingual—who were thrust unprepared into the sectarian, ethnic, and tribal honeycomb of Iraq, to distinguish friends from

---

[18] Article 37 of the 1977 Additional Protocol to the Geneva Conventions.

foes (Fallows 2006). Predators and prey with orange colouring can remain nearly invisible in green-hued grassland and woodland settings, to prey and predators with achromatic vision who are unable to differentiate between green and orange. Americans in Iraq displayed the cultural equivalent of colour blindness, call it cultural blindness, meaning that they were exceptionally vulnerable to mixing up apples and oranges. They were dressed to kill, but they could be easily played for fools. Intentionally or unintentionally stoked paranoia about invisibly lurking enemies can drive young men into lashing out indiscriminately.[19] Rather than simply duping the Americans, Iraqi insurgents who attacked Americans were able to excite in US troops an indiscriminate fear of Iraqis in general. By dressing as civilians, the militants made it difficult for the Americans to tell which Iraqis were friends and which were foes. Vexed by this engineered fog of war, American forces compensated for their inability to distinguish the dangerous from the innocent by disproportionate force, dragnet arrests, and collective punishment. The effect of these indiscriminate tactics on the way average Iraqis viewed the foreign forces was not positive.

Perfidy is a kind of meta-war crime because it predictably leads the enemy to commit other war crimes. Perfidy is a matador's cape, luring the enraged target into misdirected counter-lunges and self-defeating overkill. This can serve the strategic purposes of the mimics if the dupes' failure to respect the combatant–non-combatant distinction causes them to lose the battle for hearts and minds. Combatants fighting in civilian dress can blacken the reputation of the dupe in the same way as firing rockets from civilian neighbourhoods: both tactics can provoke the duped enemy into killing civilians. Manoeuvring the blinded enemy into unintentionally killing non-combatants can demoralize the dupe's supporters and energize its adversaries. In both cases, the mimic invites the infliction of harm on the model to ruin the reputation of the dupe among the civilian population and, ideally, to spark a popular uprising against the dupe, effectively rallying the model population to the mimic's cause. Imitating the model to trick the dupe into hurting the model, thereby turning the model against the dupe, is a good example of how insurgents and terrorists can punch above their weight. Such perfidious mimicry, too, can have lasting effects even when the mimicry itself is short-lived. This is because memory of indiscriminate massacres is long-lasting. The reputation of the dupe for barbarism, for ignoring the distinction between the innocent and the guilty, will linger long after attacks by combatants dressed as non-combatants have stopped.

---

[19] This pattern illuminates Gambetta's (2005) idea of mimic versus model-via-dupe, also taken up by Heather Hamill (in this volume) in her chapter on Northern Ireland.

As discussed earlier, an unintended if foreseeable consequence of defensive mimicry is the tendency of targeted dupes to suspect models of being mimics, a pattern that often occurs in nature. Rarer in nature but common in human conflict is the mimic's deliberate use of mimicry to gain an advantage by luring the dupe into routinely abusing innocuous models as if they were dangerous mimics. That sounds quite complicated once again, but the thought is actually quite simple.

A related pattern was clearly articulated by Machiavelli in his famous chapter devoted to ruses of war: 'above all things, a general ought to endeavour to divide the enemy's strength by making him suspicious of his counsellors and confidants' for 'this must consequently weaken his army very much' (Machiavelli 1965, Book VI 173). Because the maniacal search for moles within one's own organization can be seriously debilitating and even paralyzing, each side in a conflict has a strong incentive to make the other believe that insiders occupying strategic positions are actually outsiders in disguise. In other words, the study of wartime mimicry must be supplemented or extended by a study of mimicry's simulacrum, namely ways of conning the enemy and third parties into a paranoid fear of hostile mimics where they do not in fact exist.

False-flag operations illustrate an importantly different way in which mimics can entice dupes into punishing the mimics' models. A well-known example of a false-flag operation would be, if the charges are correct, the bombing by Russian security services of apartment blocks in Moscow and two other Russian cities in 1999, killing 293 people. By attributing the bombing to Chechen terrorists (the models), the government (the mimic) stirred up outrage in the Russian public (the dupe) and therefore artificially created a wave of support for a second Chechen war (Satter 2002; Tyler 2002). In such a case, it should be noted, the mimic is deliberately manipulating the dupe into harming the model, without having any intention of harming the dupe itself, even if that was the eventual result. In a false-flag operation, group mimicry is undertaken solely to ruin the reputation of a rival group and mobilize opposition to it.[20] Strategic mimicry of this sort, too, occurs in human society but presumably not in nature.

Spoiler Mimicry

The principal exit strategy devised by the American military in Iraq revolved around standing up an Iraqi military capable of fighting Islamic militants

---

[20] See again Heather Hamill's contribution to this volume.

and holding the country together. A wrench was tossed into this strategy by green-on-blue killings, when a native policeman or soldier working with or being trained by American-led forces lethally attacks his putative foreign allies or trainers.[21] Training can impart skill but not instil loyalty, especially not loyalty to a neutral, inclusive, non-sectarian state that did not exist outside the fantasies of the occupying power. As they were about to murder their trainers, the treacherous trainees presumably had to mimic at least briefly the behaviour of blameless trainees. That some of them may even have belonged to sleeper cells who had originally joined the Iraqi military and police with nefarious purposes was suggested by the way a 2005 Pentagon report to Congress used the term *infiltration*:

> Some insurgent infiltration of ISF [Iraqi Security Forces] undoubtedly occurs, both through the recruitment process and through bribery and intimidation. Although it is reasonable to believe that it would be more prevalent in Sunni-majority provinces, the precise extent of such infiltration cannot be known ... Because the police are often recruited by local police chiefs with little Coalition oversight, infiltration tends to be somewhat higher in the police than in the military and paramilitary forces ... The exact extent of insurgent infiltration is unknown at this time (The Pentagon 2005, 33 and 37).

One consequence of repeated green-on-blue shootings was that the American soldiers and off-duty policemen who trained the Iraqis were 'constantly on guard against the possibility that their trainees might turn against them. Even in the police headquarters for all of western Baghdad, one of the safest police buildings in the capital, the training team will not remove their body armor or helmets. An armed soldier is assigned to protect each trainer' (Paley 2006). This was not a favourable context for either training or being trained.

In southern Iraq, during this period, the British had more or less the same experience. Blame was routinely laid on Coalition efforts to stand up an Iraqi army in double time to permit an accelerated withdrawal of foreign troops. This haste necessarily entailed sloppy vetting:

> Thousands of Iraqis were recruited into the corrupt Iraqi police force, with the bare minimum of personal checking by the British military, because of the perceived

---

[21] These killings have been common in Afghanistan as well as Iraq (Ahmed 2013; M. Rosenberg 2012). Here is a report about the murder of five British soldiers at the hands of an Afghan policeman with whom they had been working closely: 'The attack occurred at midday on Tuesday in Helmand Province as the soldiers relaxed in the still-warm autumn sun on the roof of the joint checkpoint overlooking a shared British–Afghan compound. They were so much at ease that they had shed their body armour and helmets, never thinking that they would be attacked by one of the men they lived and worked with, said a local provincial official. Afterward, the attacker fled, setting off a manhunt'(Rubin, Burns, and Shah 2009).

urgency of training locals to handle security in the city. As a result, Iraqis with 100 per cent loyalty to extremist militia groups joined the police force and proceeded to use the information they gleaned from the inside to launch roadside bomb attacks on known British patrol routes (Evans 2009).

Although the ratio of bad-faith mimics to good-faith recruits was presumably very low, the fear that occasional green-on-blue killings created was psychologically decisive. The long-term consequence of even low levels of aggressive mimicry was to poison relations between foreign and Iraqi troops indelibly.

The natural reaction to rare but repeated green-on-blue killings was for foreign trainers to take an arms-length approach to all recruits. This is how identity mimicry threw a serious wrench into America's entire nation-building scheme:

> Militia infiltration of Iraq's security forces is so bad in some places that American soldiers sometimes do not know whether to trust their Iraqi counterparts. 'We don't trust 'em'", said 1st Lt. Steve Taylor, serving at a joint Iraqi-American security station in Sulakh. 'There's no way to know who's good and who's bad, so we have to assume they're all bad, unfortunately' (Katulis 2007).

The paranoid fear that every apparently good-faith collaborator is a would-be assassin in disguise, and the subsequent witch-hunt for impersonators and infiltrators, known to destroy the effectiveness of national security organizations, goes a long way towards explaining the failure of the US mission in Iraq. Strategic mimicry can corrode and potentially destroy the interpersonal trust necessary for mutually beneficial social cooperation. This is perhaps its most far-reaching effect.

In a recent green-on-blue attack in Afghanistan, an American general was among those killed. The Afghan Defense Ministry 'described the attacker as "wearing Afghan National Army uniform", which has long been a standard description offered after Afghan troops attack their foreign counterparts'. The 'standard description' here feels vaguely euphemistic, as if an unpleasant truth needed to be fogged over: 'Afghan and American commanders have said that they believed most of the insider attacks that had taken place were the work of ordinary soldiers who had grown alienated and angry over the continued presence of foreign troops here, and not carried out by Taliban fighters planted in Afghan units' (M. Rosenberg and Kakaraug 2014). To announce that a green-on-blue killer was 'wearing an Afghan National Army uniform' is nevertheless to put some verbal distance between ordinary Afghan soldiers and bad apples who go on homicidal anti-American rampages. Obfuscating the resentment of American troops among ordinary soldiers is apparently a shared interest of the Afghan and US governments. How intimations of mimicry can deflect blame from real to imaginary perpetrators brings us to our next theme.

Mimicry's Mirage

Even when they are not acting as mimics themselves, violent aggressors can use mimicry to escape retaliation. For instance, murderers can posthumously alter the appearance of the murder victim to make the death seem accidental or the killing appear justified. In Basra, during this very period, after killing a woman for sectarian reasons, the murderers would 'dress her in indecent clothes so as to justify their horrible crimes' (Mahmoud and Lanchin 2007).[22]

More relevant to our theme is the way aggressors can propagate false allegations of mimicry to establish plausible deniability. Here we encounter a predatory strategy which, once again, seems to have no analogy in the animal world: mimicry's doppelgänger. For example, if Shia policemen kidnap and murder a Sunni politician, they can muddy the waters by suggesting publicly that shadowy militiamen wearing stolen police uniforms committed the crime. This mimicry of mimicry became increasingly common towards the end of the 2006–2008 Iraqi civil war. Mimicry of the first order, militiamen passing as policemen, lent plausibility to mimicry of the second order, policemen passing as militiamen passing as policemen. And indeed the latter would have been a senseless tactic had the former not already been a widely acknowledged ploy. During that period, a series of Iraqi Interior Ministers 'repeatedly suggested that killings by gunmen in police uniforms were being carried out by impostors' (Knickmeyer 2006)[23]. Local police commanders, too, began alleging that such 'attacks are carried out by impostors who buy, steal or counterfeit police uniforms, which are freely available in many Baghdad markets' (AFP 2006). Towards the end of 2007, fending off questions about massacres committed by men dressed in military uniforms, an Interior Ministry official explained: 'Surely, they are outlaw insurgents. As for the military uniform, they can be bought from many shops in Baghdad', adding that 'we have several police and army vehicles stolen, and they can be used in the raids' (CNN 2005). Disingenuous claims of mimicry are an easy

---

[22] Another example of second hand or vicarious mimicry comes from Colombia where 'soldiers' 'are rewarded with cash or holidays for killing guerrillas but, essentially, any corpse will do in this dirty war ... In October 2008, 11 young men were enticed away from their homes in Soacha, a poor suburb of Bogotá, and offered work. A few weeks later, they were found in a grave in Ocaña, near the border of Venezuela, dressed in Farc uniforms and presented as dead guerrillas'(Power 2011).

[23] Citing both Bolani and Jabr.

way to shift responsibility for one's own criminal behaviour onto others who, because wholly imaginary, will not come forward to escape punishment by proving the inaccuracy of the charges. It is a clear example of what has come to be called "the liar's dividend."

Obscured by such politically expedient reports of mimicry was the massive infiltration of the Iraqi national police force by Shia militia. Numerous contemporary accounts testify that 'the security forces themselves have been heavily infiltrated' (*Guardian* 2006). In particular, the Ministry of Interior was 'well-known for its infiltration by Shia militias from the Mahdi army and other groups' (Tran 2007). American authorities may have been poor students of Iraqi culture, but they were not duped in this case: 'U.S. military reports on the Iraqi police often read like a who's who of the two main militias in Iraq: the Mahdi Army, also known as Jaish al-Mahdi or JAM, and the Badr Organization, also known as the Badr Brigade or Badr Corps' (Paley 2006). Summarizing these developments, an acute observer of the Iraqi government at the time remarked: 'Its militias are not infiltrators, they are an integral element of the elected parties' (Stewart 2005).[24]

Three short years after Saddam's fall, and across large swathes of Iraq, 'Shia militias had become the Iraqi police and the Iraqi army' (Rosen 2006). After the British evacuated Basra, for instance, 'the Iraqi security forces' were 'largely run by the Shia militias' (Cockburn 2007). In essence, 'the death squads became official. The Badr Brigade and the Mahdi Army, the two big Shiite militias, just joined the police forces of the Shiite-led government' (Filkins 2008, 321). As if mocking the Bush administration's project of promoting democracy, this Shia takeover of the army and police was ratified at the ballot box: 'After the elections in January 2005, the Shiite hardliners who had taken power stuffed the ministries with their own gunmen, gave them uniforms and identification cards, and turned them loose' (Filkins 2008, 316). They created what could reasonably be called *death-squad democracy*.

The Shia electoral majority guaranteed that killers inside the police would have 'political protection' in the Shia-dominated political system (Katulis 2007). When the occasional patriotic policeman tried to rein in the Mahdi Army, 'the group could pull strings in the parliament and government' to warn him off (Parker 2007). In November 2006:

[A]n Iraqi commander and four staff officers responsible for the Hurriya district were arrested on suspicion of murder, extortion and links with the Mahdi Army. The judge released them after seven days when no evidence was presented. The day

---

[24] That little has changed in the interim, despite well-publicized gestures toward formal 'inclusiveness', is suggested by a well-informed commentator's claim that in 2014 Iraq: 'Lawless Shia militias, answerable only to their leaders, supplant the army and police' (Matthews 2014).

they were released, an Iraqi lieutenant colonel who had filed a statement against the five was killed at a checkpoint (Parker 2007).

Although the checkpoint here was the government's, it served not to protect a non-denominational Iraqi state from sectarian lawbreakers, but to protect the powerful Mahdi Army from brave attempts by non-sectarian Iraqi officials to enforce the law.

As stories about police impersonators proliferated, Shia death squads, rather than imitating policemen by wearing stolen or cloned uniforms, simply joined the police force en masse and wore the uniforms that their new jobs required. Having become policemen, they were able to commit their atrocities on the job as well as after hours. By 2005, American-trained police units, filled with Shia militiamen, had started 'swooping down into Sunni neighborhoods and killing civilians and kidnapping them' (Filkins 2008, 120).

For their part, Iraq's Sunnis were 'pushed to the side, dismissed from the security forces' and, as a consequence, 'Iraq's new security forces ... were filled with young Shia' (Rosen 2006). No mimicry was involved in this case, only infiltration in the sense of reoccupation of spaces vacated by former regime loyalists. As this process continued, Shias 'gained firm control of government ministries and local police' (Rosen 2009c). The police that operated as Shia public-sector murder gangs also turned a blind eye to atrocities committed by Shia bands that continued to operate privately. Thus, the Shia militiamen who had not joined the police force had no reason to dress like policemen. This was all the more true because a police uniform was no longer likely to gull politically important Sunnis into lowering their guard.

Just as Iraq's Sunnis knew the score, so American government reports about the period are exceptionally clear-eyed, if not cynical. Here are two typical passages from a gruesome State Department study of the sectarian takeover of Iraqi state institutions released during this period:

Unauthorized government agent involvement in extrajudicial killings throughout the country was widely reported. Some police units acted as 'death squads' ... There were allegations that in May MOI [Ministry of Interior] First Division National Police officers committed extrajudicial killings of civilians in Baghdad while operating outside their duty area ... Particularly in the central and southern parts of the country, Shi'a militias—the JAM [Jaish al-Mahdi] and the Badr Organization of the Islamic Supreme Council of Iraq (ISCI)—used their positions in the ISF [Iraqi Security Forces] to pursue sectarian agendas ... on May 4, ISF members reportedly arrested and shot 14 civilians in the Jihad neighborhood. According to local residents, on May 3, personnel wearing MOI police uniforms reportedly arrested and killed 16 individuals in the Hay al-Amel neighborhood ... On April 28, individuals wearing Iraqi army uniforms reportedly arrested 31 men in the Adhamiya neighborhood; five were found dead the next day in the Kesra District (US DOS 2007).

In this passage, note that Iraqi police or army uniforms were being worn not by impostors but by newly recruited but 'bona fide' members of the Iraqi army and police. The terrifying consequence of sectarian state-capture comes across even more clearly in a second harrowing passage:

In February several high officials in the Ministry of Health (MOH), including Deputy Minister, Hakim al-Zamili, who were loyalists of Moqtada al-Sadr's JAM, were arrested and charged with organizing the killing of hundreds of Sunnis in Baghdad's hospitals, including patients, family members, and medical staff. Investigations found that under al-Zamili's direction, about 150 members of the MOH's protection service used ministry identification to move freely around Baghdad and using ambulances to ferry weapons, carried out hundreds of sectarian killings and kidnappings from 2005 to early 2007. They reportedly abducted and killed many Sunni patients at three major Baghdad hospitals, Al Yarmouk, Ibn al-Nafees, and Al Nur, as well as relatives who came to visit them or went to hospital morgues to recover their family member's bodies (US DOS 2007).

Although the use of ambulances for transporting weapons is a classical case of perfidy, no one was ever punished for the crime. The Shia government whose sectarian agenda he served released al-Zamili over American objections. He was eventually elected to parliament.[25]

Strategic mimicry was rampant in 2006–2008 Iraq. But strategic resort to *false allegations of mimicry* also occurred. So why did police officials, in this case, continue to insist that sectarian murders were being committed by impostors when all parties involved understood that the incorporation of murderous Shia militias into the police was a key feature of the sectarian takeover of the state underway in Baghdad? The answer has to do with the dilemma faced by the American-led occupation force. If we are to believe State Department and Pentagon reports, the US administration believed that an inclusive government, sharing political power, patronage, and oil revenue

---

[25] Iraq's Shia-led government made sure that the American charges failed to stick. Association with a murderous Shia militia functioned as a get-out-of-jail-free card: 'After a two-day trial, marred by accusations of witness intimidation, the charges were dropped and Mr. Zamili was freed after spending more than a year in American custody' (Santora and Gordon 2010). He was elected to the Iraqi parliament in 2010 and became a member of the security committee in 2011.

fairly among Iraq's three groups, was the only viable pathway to the kind of political stability that would allow a peaceful withdrawal of foreign troops. American hostility to a sectarian Shia takeover of the Iraqi government was evident, for example, in the approach taken to standing up the Iraqi national army. As a spokesman for the 25th Infantry Division told reporters: 'When the soldiers join the IA [Iraqi Army], they are taught in training and in day-to-day regimen of being a soldier that sectarian lines are not for the army. They are an army of one, if you will, for one nation' (Lasseter 2007). Unfortunately, the expectation of non-sectarian and multi-ethnic military and police forces serving a unified but multi-communal nation proved as unrealistic as the hope that Iraq's Shia would welcome the rearming and regrouping of Iraq's temporarily debilitated Sunnis. Unable to orchestrate the creation of an inclusive Iraqi state but desperate to create some sort of stable government in Baghdad capable of ruling the country after an American withdrawal, the US ultimately accepted or even abetted a kind of system-level mimicry. It agreed, as a price of doing business, to treat a largely sectarian government as if it were a non-sectarian national government. The claim that police impostors, not the police themselves, were perpetrating savage atrocities, was known to be fraudulent by American authorities on the ground. But American observers, many of whom were personally without illusions, nevertheless *pretended to be dupes* in order to misleadingly describe America's dismantling of the Iraqi nation as an act of nation-building to a dimly informed audience back home.

This, of course, was not the end of the story. By 2007, in response to the Shia takeover of the Baghdad government, violent Sunni jihadists, known as AQI (Al Qaeda in Iraq), had arrived with a vengeance in the Sunni Triangle. American observers were particularly worried about a marriage of convenience, based on a shared hatred of the Shias, between foreign jihadists and local Sunni tribesmen affiliated with former Baathists.[26] This development was seen to be a direct threat to the United States. General David Petraeus therefore elaborated a plan for peeling the Sunni tribal groups in Anbar province away from their co-sectarian foreign jihadists. He invited them to fight against the non-Iraqi jihadists, with whom they had plenty of differences in any case, in exchange for American money, equipment, and support. The result of long-brewing American frustration with the lack of inclusiveness exhibited by the Maliki government, the Anbar gambit was an official attempt to bring

---

[26] It is worth noticing that, 'while some analysts believe that AQI drafts Baathist insurgents to carry out its attacks, other intelligence experts think it is the other way around'(Tilghman 2007).

the Sunnis back into the Iraqi state.[27] Petraeus essentially made the same offer to the insurgent Sunni paramilitaries in Iraq as the US had originally offered to the Kurdish Peshmerga.[28] This was 'the Awakening'. It meant that a 'new breed of Sunni warlords' was 'being paid by the US to fight al-Qaida in Iraq' (Abdul-Ahad 2007).

Maliki was not supportive. From his Shia perspective it was suicidal to deputize Sunni bandits not to mention futile to spend millions 'turning poachers into gamekeepers' (Cockburn 2007). The Shia-dominated government saw the Awakening as a Sunni ploy to dupe the Americans into providing weapons and training for a future Sunni resistance. They interpreted 'incorporation' of the Sunni tribes not as a step towards nation-building, but as a Trojan Horse operation. It was an invitation to a masquerade. The Sunnis would impersonate loyal members of the Iraqi nation so long as it took to rearm themselves to confront the Shias and fight to reseize control of the Iraqi state. Shia fear of aggressive mimicry by Sunnis was made quite explicit:

> Gen. David H. Petraeus and other top commanders have hailed the initiative to enlist Iraqi tribes and former insurgents in the battle against extremist groups, but leaders of Iraq's Shiite-dominated government have feared that the local fighters known as 'volunteers'—more than 80 percent of whom are Sunni—could eventually mount an armed opposition, Iraqi and U.S. officials said (Partlow and Tyson 2007).

Thus, 'The Sunni tribesmen collaborating with American forces are not accepted by the central government in Baghdad, dominated by Shiites: The Iraqi government so far has balked at permanently hiring large numbers of the volunteers. Indeed, last month [October 2007], the Shiite political alliance of Prime Minister Nouri al-Maliki called on the U.S. military to halt its recruitment of Sunnis' (Partlow and Tyson 2007). Maliki feared that the Awakening was merely a front for anti-Shia insurgents who, once foreign forces had withdrawn, would use American training and weaponry to confront their Shia enemies. The reasonableness of this fear was confirmed by some Sunnis themselves:

---

[27] One reason why it took so long to initiate the integration strategy in such areas was apparently its incompatibility with 'democratization', namely, the ideological and historically illiterate belief that a modern democracy had no place for tribal (i.e. unelected) leaders.

[28] No one who has studied the subject says that the Peshmerga 'infiltrated' the security forces in the north. The same can be said in the Shiite areas: 'As with Shiite militias in Baghdad, the line between militia members and Iraqi security troops in Kirkuk' was 'so thin that it at times doesn't exist' (Lasseter 2007).

A senior Sunni sheikh, whose tribe is joining the new alliance with the Americans against al-Qaida, told me in Beirut that it was a simple equation for him. 'It's just a way to get arms, and to be a legalized security force to be able to stand against Shia militias and to prevent the Iraqi army and police from entering their areas', he said. 'The Americans lost hope with an Iraqi government that is both sectarian and dominated by militias, so they are paying for locals to fight al-Qaida. It will create a series of warlords. It's like someone who brought cats to fight rats, found himself with too many cats and brought dogs to fight the cats. Now they need elephants' (Abdul-Ahad 2007).

The same Sunni tribal chief concluded: 'After we finish with al-Qaida here, we will turn toward our main enemy, the Shia militias' (Abdul-Ahad 2007). And here is Seymour Hersh's report of a conversation, along similar lines, with an informed observer named Nasr:

The American policy of supporting the Sunnis in western Iraq is making the Shia leadership very nervous, Nasr said. 'The White House makes it seem as if the Shia were afraid only of Al Qaeda—but they are afraid of the Sunni tribesmen we are arming. The Shia attitude is "So what if you're getting rid of Al Qaeda?" The problem of Sunni resistance is still there. The Americans believe they can distinguish between good and bad insurgents, but the Shia don't share that distinction. For the Shia, they are all one adversary' (Hersh 2007).

Petraeus not only wanted the good insurgents to help defeat the bad insurgents. He seems to have viewed the American deal with the alienated Sunni tribesmen, quite naively, as a step toward a multi-denominational Iraqi state with a multi-denominational army. After overthrowing Saddam, the Americans adopted a co-optation strategy towards the emerging Shia fighting groups, in the hopes of converting them from lawbreakers into law enforcers. Behind this co-optation strategy lay the idea that it was better to have the Shia militias inside shooting out, rather than outside shooting in. The coalition forces had acquiesced in the Shia takeover of the army and the police in part because, as an American diplomat said: 'a complete purging of the ministry's most criminally violent employees is impossible', and because purging them from the police would not calm the situation since, if peremptorily expelled, 'they're going to go straight to the militias, or set up their own criminal gangs' (Wong and von Zielbauer 2006). The militias were officially (that is, not really) 'dissolved' in June 2004.[29] But it was clear by then that

[29] 'Coalition Provisional Authority Order Number 91: Regulation of Armed Forces and Militias within Iraq'(2 June 2004).

the military occupier lacked the power to enforce such a demob order. As a result, the Americans acquiesced in the incorporation of the Shia militias into Iraq's national security bureaucracy *faute de mieux*: 'Armed groups across Iraq reacted to the 2004 measure by enlisting in the army and police and maintaining large contingents of stand-alone militia groups, making them significantly more powerful'. It was 'a plan to discipline the militias by putting them in uniform' (Lasseter 2007). But because it favoured the Shia so one-sidedly, it ended up unleashing the militias' fury on their sectarian foe and thereby exciting a Sunni counter-reaction.

The Awakening was not only an opportunistic measure adopted to counter AQI. It was also an attempt to redress the imbalance created when armed Shia groups colonized Iraqi state structures. By accepting Petraeus's invitation to the masquerade, paradoxically, the Sunni insurgents were able to rearm and regroup under the terms of Petraeus's bargain. But the deal also had a downside, well described by Rosen:

In September 2008 Maliki—in a concession to the Americans—issued an order calling for the integration of 20 per cent of the eligible Awakening men into the ministries of defence and interior. The following month the government of Iraq began to assume responsibility (financial and otherwise) over the Awakening groups. But as of today less than five per cent have joined the Iraqi Security Forces. At the same time, senior Awakening leaders and many of their men have been arrested, while others have been relieved of their duties (and their pay) and told to go home. It is a quiet and slow process, but one that continues to emasculate one of the last groups that rivaled the authority of the Iraqi state. There is nothing the Awakening groups can do. As guerrillas and insurgents they were only effective when they operated covertly, underground, blending in among a Sunni population that has now mostly been dispersed. Now the former resistance fighters-turned-paid guards are publicly known, and their names, addresses and biometric data are in the hands of American and Iraqi forces. They cannot return to an underground that has been cleared, and they still face the wrath of radical Sunnis who view them as traitors (Rosen 2009b).

By agreeing, however opportunistically, to join forces with the Americans, the 'Sunni Awakening' insurgents lost a valuable asset, namely their anonymity and ability to elude identification. Once their names and addresses were known, their days of flying under the radar were over: 'The remaining Awakening men have burnt their bridges with their more radical former allies and are now hunted by them; the Iraqi Security Forces have improved their intelligence and strike capability and have little problem tracking those men they want to arrest' (Rosen 2009b).

Having suffered the consequences of turning against AQI in 2007–2008, the Sunni tribesmen made a different choice in 2013–2014. Taking advantage of the Syrian conflict next door, the power vacuum produced by the American withdrawal, and a renewed Saudi policy of funding proxies to combat the influence of Iran in the Arab world (Kirkpatrick 2014), the previously demoralized Sunni tribesmen joined forces with radical Islamic militants in a shocking irredentist surge (al-Salhy and Arangojune 2014). Interestingly, the same blurring of lines that characterized the relation between Sunni tribesmen and AQI before the Awakening[30] seems to characterize the current relation between IS (Islamic State) insurgents and the Sunni tribesmen. Reporting on the eighty Sunni Arab tribes plus another militant insurgent group (the Naqshbandis) consisting of former members of the Baath party with strong roots in the community who have joined IS/ISIS in the fight against Shiite power, one commentator even mentions the 'tendency of Iraqi Baath loyalists to operate through fronts' (Hassan 2014). That possibility, too, confirms the ubiquity of mimicry in Iraq's civil war.

Conclusion

Mimicry is common in wars, perhaps especially in civil wars. Members of groups that are commonly victims of lethal attack naturally strive to resemble members of groups less likely to be attacked. And although prohibited from doing so by international law, belligerents continue to wear enemy uniforms or civilian attire when launching surprise attacks. The tactical advantages gained by such perfidy are apparently too great to resist. This is just as true of Iraq as elsewhere. But many of the reported mimicry incidents of 2006–2008 also call for a more specific account. Contextual factors, specific to Iraq during this period, undoubtedly played a role.

As murder gangs began to terrorize the country, reports of defensive mimicry flooded the press. While sectarian death squads provided the motive for identity mimicry, internal displacement created the opportunity. Reports of defensive mimicry tapered off less because detection efforts were improved than because residential segregation along sectarian lines reduced both the opportunities and the motives for impersonation. The hardened shells that grew up around sectarian groups was a natural defensive response

---

[30] See Cockburn (2007).

to inter-communal savagery. As communities became more homogeneous, moreover, members of minority sects began to stick out like sore thumbs, making identity mimicry more difficult and therefore further accelerating the move towards residential segregation.

An additional hypothesis, harder to verify but plausible on its face, is that a brief period of successful defensive mimicry provoked acute suspiciousness on the part of identity monitors which, in turn, resulted in genuine members of a sect being abused as if they were impostors. In response, these victims of mimicry's externalities naturally had an incentive to develop identity signals that, if not imitation-proof, were at least harder to imitate. That such a sharpening of sectarian identity, an immune response triggered by defensive mimicry, played a role in the intensification of sub-allegiances that virtually destroyed Iraqi national identity seems quite plausible.

Be that as it may, the story of aggressive mimicry in Iraq's sectarian conflict is different. Aggressive mimicry in wartime can survive ramped-up detection efforts because mimicry detection will never reach 100 per cent in the few moments it takes to spring an ambush. Hence, aspiring *embuscadiers* can reasonably expect some payoff from impersonating innocents or feigning membership in the target group when carrying out an attack. Uninterrupted reports of attackers wearing lookalike uniforms and driving cloned vehicles provide strong evidence that this calculation continues to be at work.

It is important, finally, that the mimic-model-dupe triad, when applied to the Iraqi civil war, is not only a social scientist's illuminating construct, but also a partisan politician's obscurantist chimera. The basic reality of the 2006–2008 period was the complicity between the Shiites in the Iraqi government, army, and police and the Shiite militias and death squads. As the Shia parties captured the Iraqi state, Shia militias not only used and were used by, but actually became, the Iraqi army and the police. It is not quite accurate to conceive of this overlap as 'infiltration', because there never was a genuinely non-sectarian army and police, loyal to a non-sectarian central government, into which sectarian forces could be infiltrated. As a result, reports filed in the Western press, stating that militiamen who had 'dressed in Iraqi police or army uniforms' while committing bloody massacres, were misleading even if literally accurate. Sometimes the killers were policemen pretending to be militiamen pretending to be policemen.

Calculated feigning can be calculatingly feigned. This faking of faking, or imitation squared, functioned as a readily available strategy for deflecting blame and fostering the illusion that the Shiites, taking over all the organs of the Iraqi state under the noses of the Americans, were themselves non-partisan and uninterested in monopolizing power. This was a useful illusion for Maliki's government to feed the US occupiers. It was a kind of system-level mimicry. The Shia power sought to convince the American paymasters that it was at least trying to become an inclusive Iraqi state. The aggressive mimicry to which the Iraqi civil war gave rise was real enough. But its simulacrum was also taken up and deployed to hide the dark complicity of the Baghdad government with coreligionist murder gangs. By 2008, if not before, the Iraqi state 'belonged to the Shias' (Rosen 2009c). After the Shiites had consolidated their take-over of the Iraqi state, they no longer had much need to resort to false allegations of identity mimicry, which may be why we observe a slight decrease in press reports of aggressive mimicry, too, after Maliki's March 2008 attack on the Shiite militias in Basra. Rates of aggressive mimicry by Sunni insurgents, by contrast, did not decline. What declined was the pretense by the Maliki government that Shiite militias, when committing atrocities, were impersonating agents of the state. Especially after the American withdrawal in 2009, the Shiite-led government in Baghdad was able to discard this particularly macabre disguise.

References

Abdul-Ahad, Ghaith. 2007. 'Meet Abu Abed: The US's New Ally against Al-Qaida'. *Guardian*, 10 November.

AFP. 2006. 'Iraqi Police Don New Uniform to Beat Impostors'. *AFP*, 10 October.

Ahmed, Azam. 2013. 'In Insider Attack, an Afghan Soldier Opens Fire on Coalition Forces, Killing One'. *New York Times*, 9 July.

Ali, Ahmad and Oliver Poole. 2006. 'Sunnis Learn Shia Customs to Bluff Baghdad Death Squads'. *Telegraph*, 9 October.

Aljazeera. 2010. 'Iraq Targets Sale of Fake Uniforms'. *Aljazeera*, 16 February.

al-Salhy, Saudad and Tim Arangojune. 2014. 'Sunni Militants Drive Iraqi Army Out of Mosul'. *New York Times*, 10 June.

AP. 2006. 'Gunmen Ambush Minibuses South of Baghdad'. *AP*, 11 November.

Bacon, Francis. 1985. 'On Simulation and Dissimulation'. In *The Essays*. Hammondsworth: Penguin.

Baker, Peter. 2003. 'Iraqi Man Risked All to Help Free American Soldier'. *Washington Post*, 4 April.

Barnard, Anne and Jodi Rudoren. 2014. 'Despite Israeli Push in Gaza, Hamas Fighters Slip Through Tunnels'. *New York Times*, 19 July.

Buckley, Cara. 2007. 'Gunmen Dressed as Iraqi Troops Kill at Least 11 in Village Near Baghdad'. *New York Times*, 23 November.

Burns, John F. and John O'Neil. 2006. 'Iraqi Government Declares State of Emergency'. *New York Times*, 23 June.

Byers, Michael. 2006. *War Law: Understanding International Law and Armed Conflicts*. New York: Grove Press.

Cave, Daniel. 2007. 'Death Toll in Iraq Bombings Rises to 250'. *New York Times*, 15 August.

CNN. 2005. 'Sunni Sheik, Family Members Slain'. *CNN*, 24 November.

Cockburn, Patrick. 2007. 'Ignominious End to Futile Exercise That Cost the UK 168 Lives'. *Independent*, 3 September.

Cockburn, Patrick. 2008. 'Iraq: Violence Is down—but Not because of America's "Surge"'. *Independent*, 14 September.

Dillon, Dana R. 2003. 'Perfidy in Iraq. Their Tactics, Our Response'. *National Review Online*, 26 March.

Erlanger, Steven. 2009. 'A Gaza War Full of Traps and Trickery'. *New York Times*, 10 January.

Evans, Michael. 2009. 'Insurgents Will Have Learnt from Vetting Failures in Iraq How to Infiltrate Security'. *The Times*, 5 November.

Fallows, James M. 2006. *Blind into Baghdad: America's War in Iraq*. New York: Vintage Books.

Filkins, Dexter. 2008. *The Forever War*. New York: Knopf.

Gambetta, Diego. 2005. 'Deceptive Mimicry in Humans'. In *Perspectives on Imitation: From Neuroscience to Social Science*. Volume 2: *Imitation, Human Development and Culture*, edited by Susan L. Hurley and Nick Chater, 221–242. Cambridge, MA: MIT Press.

Gardner, David. 2015. 'Ramadi Takeover Is yet Another Symbol of a Hollowed-out State'. *Financial Times*, 18 May.

Gordon, Michael R. 2012. 'Against U.S. Wishes, Iraq Releases Man Accused of Killing American Soldiers'. *New York Times*, 16 November.

Guardian. 2006. 'Dangerous Impostors!'. *Guardian*, 14 November.

Harding, Luke. 2014. 'Iraq Suffers Its Deadliest Year since 2008'. *Guardian*, 1 January.

Hassan, Hassan. 2014. 'More than ISIS, Iraq's Sunni Insurgency'. *SADA, Carnegie Endowment for International Peace*, 17 June.

Hersh, Seymour M. 2007. 'Shifting Targets: The Administration's Plan for Iran'. *The New Yorker*, 8 October.

Hider, James and Ali al-Hamdani. 2006. 'Faking Faith to Fool Death Squads: For Sunnis, Being Able to Pass as a Shia Could Save Their Lives'. *The Times*, 15 August.

Hobbes, Thomas. 1968. *Leviathan*. Harmondsworth: Penguin.

Katulis, Brian. 2007. 'Killing the Patient: Iraq's Security Forces Are Part of the Problem'. *Center for American Progress*, 11 June.

Khaldoun, Ibn. 1863. *Les Prolégomènes* (trans. William MacGuckin de Slane). Paris: Librairie orientaliste Paul Geuthner.

Khedery, Ali. 2014. 'Why We Stuck with Maliki—and Lost Iraq'. *Washington Post*, 3 July.

Kirkpatrick, David D. 2014. 'Power Struggles in Middle East Exploit Islam's Ancient Sectarian Rift'. *New York Times*, 5 July.

Knickmeyer, Ellen. 2006. 'Official: Guard Force Is Behind Death Squads'. *Washington Post*, 14 October.

Lasseter, Tom. 2007. 'In Iraq, Kurdish Militia Has the Run of Oil-Rich Kirkuk'. *McClatchy Newspapers*, 16 February.

Lattimore, Richmond and David Grene (eds). 1958. *Euripides IV*. Chicago, IL: University of Chicago Press.

Linderman, Gerald F. 1999. *The World within War: America's Combat Experience in World War II*. Cambridge, MA: Harvard University Press.

Lucas, James Sidney. 1985. *Kommando: German Special Forces of World War Two*. New York: St. Martin's Press.

Lucas, Ryan. 2014. 'Iraq Violence Claimed More Than 2,400 Lives In June, UN Reports'. *Huffington Post*, 1 July.

Machiavelli, Niccolò. 1965. *The Art of War*. Cambridge, MA: Da Capo.

Machiavelli, Niccolò. 1996. *Discourses on Livy*. Trans. Harvey Mansfield and Nathan Tarkov. Chicago, IL: University of Chicago Press.

Mahmoud, Mona and Mike Lanchin. 2007. 'Basra Militants Targeting Women'. *BBC World*, November.

Matthews, Jessica T. 2014. 'Iraq Illusions'. *New York Review of Books*, 14 August.

McNeil, Donald G. Jr. 2012. 'C.I.A. Vaccine Ruse May Have Harmed the War on Polio'. *New York Times*, 9 July.

Moir, Lindsay. 2009. 'Conduct of Hostilities—War Crimes'. In *The Legal Regime of the International Criminal Court*, edited by Jose Doria, Hans-Peter Gasser, and M. Cherif Bassiouni, 487–536. Leiden: Brill.

New York Times. 1947. 'Court Holds Former SS Officer and Seven Aides Did Not Violate the Rules of War During Battle of Bulge'. *New York Times*, 10 September.

New York Times. 1989. 'Disguised Israelis Blamed in Slaying'. *New York Times*, 20 August.

Oakford, Samuel. 2018. 'Counting the Dead in Mosul: The civilian death toll in the fight against ISIS is far higher than official estimates'. *The Atlantic*, 5 April.

Packer, George. 2014. 'A Friend Flees the Horror of ISIS'. *The New Yorker*, 6 August.

Paley, Amit R. 2006. 'In Baghdad, a Force Under the Militias' Sway: Infiltration of Iraqi Police Could Delay Handover of Control for Years, U.S. Trainers Suggest'. *Washington Post*, 31 October.

Parker, Ned. 2007. 'Shiite Militia Infiltrates Iraqi Forces'. *Los Angeles Times*, 16 August.

Parks, W. Hays. 2003. 'Special Forces' Wear of Non-Standard Uniforms'. *Chicago Journal of International Law* 4 (2): 493–560.

Partlow, Joshua and Ann Scott Tyson. 2007. 'Hurdles Stall Plan For Iraqi Recruits: Shiite Leadership Wary of Bringing Fighters Into Ranks'. *Washington Post*, 12 November.

Pentagon. 2005. 'Measuring Stability and Security in Iraq'. Report to Congress.

Persico, Joseph E. 1979. *Piercing the Reich: The Penetration of Nazi Germany by American Secret Agents during World War II*. New York: Viking Press.

Power, Mike. 2011. 'The Devastation of Colombia's Civil War'. *Guardian*, 23 April 23.

Rasheed, Ahmed and Kareem Raheem. 2014. '50 Killed in Bomb Attack on Rally, Police and Troops Voting in Iraq'. *Reuters*, 28 April.

Reid, Robert H. 2007. 'Gunmen Dressed as Iraqi Soldiers Kill 15'. *AP*, 19 May.

Reuters. 2014. 'Kurds Buy Arms, Fear Infiltration as Islamic State Eyes Arbil'. *Reuters*, 9 August.

Rhode, David. n.d. 'Perfidy and Treachery'. *Crimes of War*. http://www.crimesofwar. org/a-z-guide/perfidy-and-treachery.

Rosen, Nir. 2006. 'Anatomy of a Civil War'. *Boston Review*, December.

Rosen, Nir. 2007. 'Security Contractors: Riding Shotgun With Our Shadow Army in Iraq'. *Mother Jones*, 24 April.

Rosen, Nir. 2009a. 'The Gathering Storm'. *The National*, 10 April.

Rosen, Nir. 2009b. 'The Big Sleep'. *The National*, 24 April.

Rosen, Nir. 2009c. 'An Ugly Peace: What Changed in Iraq'. *Boston Review*, December.

Rosenberg, Carol. 2014. 'Guantanamo Detainee Abd Al Hadi Al Iraqi Charged With "Perfidy"'. *Miami Herald*, 18 June.

Rosenberg, Carol. 2022. 'Commander of Afghan Insurgency Pleads Guilty at Guantánamo Bay'. *New York Times*, 13 June.

Rosenberg, Matthew. 2012. 'Training Afghan Allies, With Guard Firmly Up'. *New York Times*, 25 September.

Rosenberg, Matthew and Haris Kakaraug. 2014. 'U.S. General Is Killed in Attack at Afghan Base, Officials Say'. *New York Times*, 5 August.

Rozenberg, Joshua. 2003. 'The Perils of Perfidy in Wartime'. *Telegraph*, 3 April.

Rubin, Alissa J. 2009. 'From Iraq, Lessons for the Next War'. *New York Times*, 1 November.

Rubin, Alissa J., John F. Burns, and Taimoor Shah. 2009. 'Troop Deaths in Afghanistan Stir Outcry in Britain'. *New York Times*, 5 November.

Santora, Marc and Michael R. Gordon. 2010. 'Murky Candidacy Stokes Iraq's Sectarian Fears'. *New York Times*, 3 March.

Satter, David. 2002. *The Shadow of Ryazan: Who Was Behind the Strange Russian Apartment Bombings in September 1999?* Washington, DC: The Hudson Institute, Johns Hopkins School of Advance International Studies. http://www.wanttoknow. info/documents/false_flag_russia_bombings.pdf.

Savage, Charlie. 2012. 'Prisoner in Iraq Tied to Hezbollah Faces U.S. Military Charges'. *New York Times*, 23 February.

Scientific American. 2013. 'How the CIA's Fake Vaccination Campaign Endangers Us All'. *Scientific American* 308 (5): n.p.

Sieff, Kevin and Sayed Salahuddin. 2014. 'Attack on Base in Afghanistan Leaves at Least One Member of U.S.-Led Coalition Forces Dead'. *Washington Post*, 20 January.

Sophocles. 1953. In *Electra and Other Plays*. London and New York: Penguin.

Spencer, Richard. 2014. 'What It Feels like to Be on the Receiving End of Isis's Pickup Truck Killing Party'. *Telegraph*, 28 June.

Spinner, Jackie. 2004. 'Head Scarves Now a Protective Accessory in Iraq: Fearing for Their Safety, Muslim and Christian Women Alike Cover Up Before They Go Out'. *Washington Post*, 20 December.

Steele, Jonathan. 2008. *Defeat: Why America and Britain Lost Iraq*. Berkeley, CA: Counterpoint.

Stewart, Rory. 2005. 'Losing the South'. *Prospect Magazine*, November.

Straus, Ulrich. 2003. *The Anguish of Surrender: Japanese POW's of World War II*. Seattle, WA: University of Washington Press.

Tavernise, Sabrina. 2005. 'British Army Storms Basra Jail to Free 2 Soldiers From Arrest'. *New York Times*, 20 September.

Tavernise, Sabrina. 2006. 'Gunmen in Iraqi Army Uniforms Kidnap a Wealthy Banker and His Son'. *New York Times*, 18 February.

Tavernise, Sabrina and Karim Hilmi. 2007. 'An Oasis from Politics amid the Turmoil in Baghdad'. *International Herald Tribune*, 13 November.

Tilghman, Andrew. 2007. 'The Myth of AQI'. *The Washington Monthly*, October.

Tran, Mark. 2007. 'Iraqi Police Cannot Control Crime'. *Guardian*, 30 May.

Tyler, Patrick E. 2002. 'Russian Says Kremlin Faked "Terror Attacks"'. *New York Times*, 1 February.

Tzu, Sun. 2006. *The Art of War*. New York: Penguin.

US DOS. 2007. 'Iraq—Country Reports on Human Rights Practices'. Washington, DC: United States Department of State, Bureau of Democracy.

Vanden, Tom and Zaid Sabah. 2006. 'Gunmen Kidnap Dozens from Iraq Ministry Office'. *USA Today*, 15 November.

Wong, Edward and Paul von Zielbauer. 2006. 'Iraq Stumbling in Bid to Purge Its Rogue Police'. *New York Times*, 17 September.

Index

For the benefit of digital users, indexed terms that span two pages (e.g., 52–53) may, on occasion, appear on only one of those pages.

A

Abbottabad 215–216
activism, Internet and 51, 80
Adair, Johnny 99–101
Adams, Gerry 105
Afghanistan, infiltration of security forces 231–232
aggressive mimicry *see under* mimicry
AIVD 61–62, 70
American Civil War 17–18
Anarchists 116, 118
ancient period, mimicry in warfare 14–16, 19–20
Anderson, Ryan 63–64
Ansar al-Mujahideen forum 50–51, 59–60, 66–67, 72, 75
Ansari, al- 13
AQI (Al Qaeda in Iraq) *see under* Qaida, al-
Artemisia I of Caria 15
artificial intelligence 9–10
Asadi, 'Ali Abdel Hussein al- 210
Ask.fm 81
assassinations 97t, 96–101, 110, 122, 122f, 123f, 127, 128f, 129f, 133; *see also* political violence
authentication 11, 111, 177, 188
 third-party authentication 187–202
 see also code words; identity signalling; vetting agencies
Autonomia Operaia 116, 132, 136, 140–141
Avanguardia Nazionale 117
avatars 56
Averill, Liam 95–96
Awakening, the 236–237, 239–240
Awlaki, Anwar al- 78
Aziz, Ziad Taha 216–217 n. 5

B

Bacon, Francis 208–209
Baghdad 205–207, 213–214

Banca dell'Agricoltura (Milan) bombing 117–119
Barak, Ehud 218
Basra 206, 232–233, 241–242
Batal forum 68
Bates, Henry W. 14
Bektasevic, Mirsad 78–79
Bersezio, Vittorio 143–144 n. 18
Biblical references 2, 19
Birmingham Mail 104–105, 107–108
bombings and hoaxes
 Italy 116–119, 122f, 123f, 127, 128f, 129f, 133
 UK 93–94, 103–104, 106–110
 see also code words; political violence
boundaries 94–95, 109, 166–171
Brigate Garibaldi 141
Brigate Rosse (BR) 116, 121
 claiming attacks 124–125, 139 n. 14, 140–141, 143
 claiming methods 138t, 138–139, 144–146
 counterclaims 134–135
 disclaimers 136, 143
 false claims against 129–130, 132, 140, 143
 name 141
 logo 142, 142f
 not claiming attacks 131
 other claiming BR attacks 132–133
 signature 143–144, 147
British Army 91–93, 99
British Transport Police 108

C

camouflage 7 n. 7, 92–94, 110, 206–209
Camp Chapman 217 n. 6
Cannon, Walter 22–23
CdS *see Corriere della Serra*
Chares 19
Chechen terrorists 229
Churchill-Coleman, George 106–107

CIA 13, 61–62, 215–216
Cicero 144–145
CIRA *see* Continuity IRA
Circolo XXII Marzo 118
civil wars 155
    bribery 162
    camouflage 206–209
    cultivating friendships 162
    cultural proximity 155–158
    displacement of population 213
    flight 160–161
    hiding 161–162
    local knowledge 166
    mimicry 3, 20–23, 41–42, 162–163,
        168–173, 213–214, 240
    see also political violence
Clausewitz, Carl von 13–14
code words 90–91, 102–109, 111, 137; *see
    also* bombing and hoaxes; identity
    signalling
Codrus, King of Athens 16
Colombia, killing of guerrillas 232 n. 22
Communist Party
    India 177–178, 183, 195, 197–198
    Italy 117, 120, 120*f*, 141
computer-mediated communication (CMC)
    55
Coniglione, Agatino 139 n. 14
Conrad, Joseph 118
Continuity IRA (CIRA) 91, 103–104
Coogan, Tim Pat 92–93
Corriere della Serra (CdS) 120–121
Costa, Pietro 131
costly signs *see under* signalling theory
criminals
    mimicry and 22, 110–111, 116, 128–131,
        170–171
    trust and 3
cultural proximity 153–158, 165–166
customs systems 41–42

D

DAAD *see* Direct Action Against Drugs
Daqduq, Ali Musa 223
Dark Web Forum Portal (DWFP) 59
Davidson, Brendan 98–99
Deaglio, Enrico 143–144 n. 18
deception, mimicry and 4–6
deceptive mimicry *see under* mimicry
deep fakes 9–10

Delle Chiaie, Stefano 117
De Pauw, Linda 17–18
Des Forges, Alison 162–163
DHK & Co Solicitors 110–111
Diani, Mario 80
Direct Action Against Drugs
    (DAAD) 102
distribution companies 76–77, 77*f*
Djehuty, General 12–13
doubles 10, 21
Dulami, Kadhim Sarheed
    Ali al- 220–221
Dutto, Attilio 139 n. 14

E

Ekhlaas, al-, forum 61–62
Electronic Mujahidin Network 68
Eliav-Feldon, Miriam 12
Ellsberg, Daniel 176
Elster, Jon 170–171, 182, 185–186
Epic of Gilgamesh 12–13
equilibrium rate *see under* mimicry
Escobar, Pablo 22 n. 15
espionage *see* spying
Ethnologue 157
Euripides 224 n. 16
extortion
    India 176, 178*t*, 177–178, 186, 189, 193
    Italy 127, 128*f*, 129*f*, 129–130, 130*f*, 136*f*
    Northern Ireland 110–111

F

Facebook 52, 55–56, 81
Fajr Media, al- 61–62, 68, 76–77
Falluja, al-, forum 54, 59–60,
    66–69, 71, 75
false flag operations 21–22, 90, 100–102,
    116–117, 127–128, 127*t*, 128*t*, 129*f*,
    131–132, 229
Fearon, James 156–157
Fida', al-, forum 54, 59–60, 75
Finkel, Evgeny 18
Fioravanti, Valerio 133–134
Floyd, Kory 55
Franceschini, Alberto 119 n. 5, 142
fraud 55–56
Freddi, Guido 132, 136
free-riders *see under* mimicry
Frontinus 13, 15–16

G

Gambetta, Diego 3, 27, 47, 56–57, 74,
    152–154, 156
GCHQ *see* Government Communication
    Headquarters
Gellner, Ernest 156
Geneva conventions 13–14, 222–223
genocide 28, 32, 45–46; *see also* Holocaust;
    Rwanda genocide
Global Islamic Media Front 76–77
Global Terrorism Database (GTD) 120–121
Gourevitch, Philip 160, 162–163, 166,
    172–173
Government Communication Headquarters
    (GCHQ) 61
Green, John Francis 95–96
Guichaoua, André 165

H

Habyarimana, James 173, 176–177
Hague Convention IV (1907) 222–223
Hamid Shifi 220
Hancock, Jeffrey 55
Harawi, al- 13
Hatzfeld, Jean 161–162
Hereward the Wake 16–17
Hersh, Seymour 238
Hirschman, Albert 158
Hisba, al-, forum 61–62, 68
Hobbes, Thomas 224
Holocaust 152–153
Horowitz, Donald 155
Hughes, Brendan 95–96
Huntington, Samuel 158

I

Ibn Khaldun 205–206, 224
identity mimicry *see* mimicry
identity signalling 6–7, 47, 102–109
    faces as 9–10
    fake IDs 43–44, 46 n. 17, 162–163
    group identity 7, 115
    ID cards 162–163, 165–168
    identity fraud 2–3
    Internet and 50–51, 55–56, 70
    noms de guerre 105
    passwords 52–53, 189
    repeat transactions 188–190
    technology as threat to 9–10
    third-party authentication 187–188

verification problems 17
    see also code words; signalling theory,
        costly signs
IDF *see* Israel Defense Forces
IHL *see* International Humanitarian Law
imposters 17
India
    caste system 169–170, 196
    political violence 177–178, 179f
    see also Naxals/Naxalites
infiltration *see under* political violence
INLA *see* Irish National Liberation Army
Inspire 62
intelligence gathering *see under* political
    violence
International Humanitarian Law (IHL)
    225–226
Internet
    activism and 51, 80
    avatars and 56
    confidence in technology 57
    forums 52–53
    romantic relationships and 55
    terrorism and 50–51
    trust and 51, 55–58, 80
    web proxies and https URLS 67
    see also jihadi Internet forums
IPLO *see* Irish People's Liberation
    Organisation
IRA *see* Irish Republican Army
Iraqi, Abd al-Hadi al- 222–223
Iraqi, Abu Maysara al- 78–79
Iraqi civil war 205
    aggressive mimicry 216–222, 241
    Awakening, the 236–237, 239–240
    defensive mimicry 213–216, 240–241
    fake ID cards 210–211
    infiltration of police force 233–234,
        238–239, 241
    infiltration of security forces 229–231,
        235, 238–239, 241
    mimicry of mimicry 232–242
    models assisting mimics 211
    models harmed by mimicry 215–216
    perfidy 222–229
    residential segregation 214–216
    Sunnis mimicking Shiias 211–213
    uniforms, deceptive wearing of 218,
        220–223, 223–224 n. 15, 241
Irish National Liberation Army (INLA) 97t

Irish People's Liberation Organisation
    (IPLO) 97*t*
Irish Republican Army (IRA) 44–45, 90–93
    assassinations 96, 97*t*, 101
    bombings and hoaxes 103–104, 106–109
    code words 105–108, 111
    criminality 182, 185–186
    false flag operations 102
    intelligence gathering on 100
    internment 93
    mimicked by other groups 103–105,
        110–111
    noms de guerre 105
    prison escapes 95–96
ISIS (IS/Islamic State) 21–22, 205–206,
    216–217 n. 5, 240
Islamist Internet forums *see* jihadi Internet
    forums
Israel Defense Forces (IDF) 218
Italy
    assassinations 122, 122*f*, 123*f*, 127, 128*f*,
        129*f*, 133
    bombings 116–119, 122*f*, 123*f*, 127, 128*f*,
        129*f*, 133
    extortion 127, 128*f*, 129*f*, 129–130, 130*f*,
        136*f*
    political violence and, 120*f*, 122*f*, 124*f*
    Second World War resistance 117, 141

J
Jaffa 12–13
Jahad forum 59–60, 75
Jews
    ethnic boundaries 169–170
    Germany Army, joining 18, 152–153
    Holocaust 152–153, 153*f*
    mimicking non-Jews 10, 45–46, 152–153
Jihad, Sayf al- 69
jihadi Internet forums 50, 53*f*
    2000s–2010s 52–55
    countermimicry 51
    data collection 58–60
    distribution companies 76–77, 77*f*
    honeypot operations 58, 61–62
    infiltration of 62–63, 66–68, 72
    knowledge, complex 74, 78–79
    media formats 74, 77–78
    media production and distribution
        entities (MPDEs) 77
    mimicry and 57, 60–64

propaganda 76–80
recruitment, no evidence of 65–66
registration 52–53
reputation systems 74–76, 75*f*
spying and 1–2, 57, 59, 66–67, 69, 71–72
surveillance and 60–61, 67
time as costly sign 73–74, 80
trust and 51, 55, 57–58, 63–71
trustworthiness, signs of 71–73
vetting agencies 74, 76–77
videos 78

K
Kagame, Paul 160
Kalyvas, Stathis 155, 160–161, 167 n. 20
Keenan, Sean 99
Khawla bint al-Azwar 17–18
kidnappings 122, 122*f*, 123*f*, 127, 128*f*,
    129*f*, 130–131, 133, 220
Kimmage, Daniel 77
Kristjánsson, Mímir 11
Kubba, Hassan 218

L
Labi, Nadya 70
Lachares 16
Lancaster, Thomas, Earl of 16–17
Lawrence, T. E. 208
Leonard, Elizabeth 17–18
Libi, Abu Yahya al- 69
Lieber Code (1863) 222–223
linguistic differences 157–158
Liu Bang 14–15
Lotta Continua 116, 123–124, 140–141
Louis VI, King of France 16–17

M
Macchiarini, Idalgo 144
McDonald, Henry 107–108
McDowell, Thomas 90
McEvoy, Kieran 103–104
McIver, John 94–95
Machiavelli, Niccolò 224, 229
McKittrick, David 96
Mafia 129–130, 176
Magee, Patrick 94
Maine, Hugh, Count of 16–17
Makki Abu Dharr, al- 79
Maliki, Nuri al- 205–206, 237, 239, 241–242
Maqdisi, Abu al-Nur al- 69

MAR *see* Minorities at Risk
Ma'sada, al-, forum 68
maskirovka 18
Masullo, Juan 39 n. 10
Mattarella, Piersanti 129–130
Maze Prison 95–96
Mechelli, Girolamo 139 n. 14
media production and distribution entities
    (MPDEs) 77
medieval period, mimicry in warfare
    16–18
Merlino, Mario 118
MI5 100
Middle East Research Institute
    (MEMRI) 61
mimicry 1–4
    adjustment 38–41, 40*f*, 41*f*
    aggressive mimicry 96–102, 109–110,
        170–171, 216–222, 241
    anti-mimicry measures *see*
        countermeasures
    as bluffing 27
    by criminals 22, 110–111, 116, 128–131,
        170–171, 176, 180–181
    by ex-models 181
    by groups 21–22, 196
    by intelligence services 61–64, 99–100
    by rival groups 100–104, 110, 115–116,
        126–128, 132
    camouflage vs. 208–209
    cities and 46 n. 16, 48
    costs 11–12, 30, 34, 38, 43, 46–47,
        153–154
    countermeasures 8–9, 12, 22, 51, 166–169,
        221–222; *see also* signalling theory,
        costly signs
    cultural proximity and 153–154, 158,
        165–166
    deceptive mimicry 6–7
    defensive mimicry 94–96, 99–100, 109,
        154, 171, 173, 213–216, 240–241
    detection 27–32, 36*f*, 38, 169, 214
    deterrence 12, 43
    dupes/targets and 6–9, 11, 27, 57
    equilibrium rate 27–28, 32, 34, 38, 42–44,
        153, 166–167, 214, 217
    exaggerating size of groups/threat 126
    fake IDs 43–44, 46 n. 17, 162–163

false flagging 21–22, 90, 100–102,
    116–117, 127–128, 127*t*, 128*t*, 129*f*,
    131–132, 139, 140*t*, 229
free-riders 110–111, 115–116, 126–127,
    127*t*, 128*t*, 129*f*, 129, 132–135, 138,
    138*t*, 139, 140*t*
honeypot operations 58, 61–62
infiltration by groups/infiltration of
    groups *see under* political violence
institutional commitments to deter 41–43
intelligence gathering and 99–100
local knowledge a barrier to 166
men passing as women 1–2, 21
mimicry of mimicry 232–242
minorities and 154–156, 158, 168–169
models and 6–7, 10–11
models assisting mimics 162, 211
models harmed by mimicry 215–216,
    225–227, 229, 241
offensive mimicry *see* aggressive mimicry
passing 28, 30–32, 35, 36*f*, 39
payoffs 27–28, 30, 31*t*, 42
population share 33–35, 37*f*, 37
prevention *see* countermeasures
props 212
purging as a result of 72
responses to 133–137
sleeper cells 216–217, 229–230
social mimicry 27
specific individuals mimicked 17, 21
stakes 28, 30–31, 33–34, 36*f*, 36–37
as strategic game 8–12, 27, 29–37
as subcategory of deception 4–6
uniforms and 218–224
wartime 12–20
women passing as men 17–18
see also spying
Minorities at Risk (MAR) 154–157, 172
Modeste, Abbé 161
Montoneros 132
Moretti, Mario 131, 145–146
Moro, Aldo 116, 145
Morrison, Danny 105
Movimento Sociale Italiano 116
MPDEs *see* media production and
    distribution entities
Muhammad 12–13
Mulholland, Joe 106
Müller, Fritz 14
Museveni, Yoweri 160

N

Nairac, Robert 100
Naji, Hamudi 215
National Security Agency (NSA) 61
Naxals/Naxalites 170–171, 176, 178*t*, 179*f*
   costly signs 190–199, 191*t*, 202–203
   fake Naxals 180–181, 185–187, 202
   intelligence gathering 201–202
   signalling, forms of 187–202
   threat-execution 183–184
   trademark 183–184, 186
Ndagijimana, Samuel 161
Nkongoli, Lurent 162
Nkurunziza, François Xavier 161
noms de guerre 105
Northern Ireland conflict 90
   assassinations/aggressive mimicry 97*t*,
      96–102, 109–110
   bombings and bomb hoaxes 93–94,
      103–104, 106–110
   camouflage 92–94, 110
   code words 90–91, 102–109, 111
   defensive mimicry 94–96, 109
   intelligence gathering 99–100
   internment 92–93
   Northern Ireland (Sentences) Act (1998)
      101–102
   prison escapes 95–96
Nuclei Armati Proletari (NAP) 116, 125,
   134–135, 139 n. 14
Nuclei Armati Rivoluzionari (NAR) 116,
   125, 133–135
Nucleo Armato Rivoluzionario Nazionale
   141–142
Nucleo Fascista Rivoluzionario 129–130

O

Obama, Barack 10
Óglaigh na hÉireann 91
Omagh bombing 107–108
Omar, Yassin 1–2
O'Neill, Daniel 94–95
O'Neill, Terence 90
Operation Aviv Neurim 218 n. 7
Operation Cupcake 62
Operation Demetrius 93
Operation Mincemeat 5
Ordine Nero 134–135
Ordine Nuovo 129–130, 134–135

P

Paltalk 78, 80–81
Pardi, Francesco 119
Parks, Malcolm 55
passing *see under* mimicry
Patriot Act (2001) 64
perfidy 206, 222–229
Peshmerga 236–237
Petraeus, David 236–239
police
   bribery and 162
   code words and 106, 108–109, 111
   infiltration by militias 233–234, 238–239,
      241
   infiltration of groups 51, 118
   mimicking by 22, 97*t*, 206, 232–233, 241
   mimicking of 97*t*, 98–99, 208, 219–222,
      229–230, 232–233
   as unable to protect civilians 184, 193
   as unlikely to report mimicry 185
Police Service of Northern Ireland (PSNI)
   91
Polish resistance 10
political violence 114, 120*f*, 122*f*, 124*f*
   camouflage 92–94, 110
   claims by groups 114–115, 122–126, 125*t*
   claiming methods 138–139, 138*t*, 140*t*
   code words 90–91, 102–109, 111, 137
   criminals mimicking 22, 110–111, 116,
      128–131, 170–171, 180–181
   disclaims and counterclaims 133–137,
      136*f*
   discrediting other groups/shifting blame
      see false flag operations
   false claims by groups 115–116, 126–133,
      127*t*, 128*t*, 128*f*, 129*f*, 130*f*, 136*f*
   false claims by groups, preventing
      137–140
   funding 65–66, 131–132, 147–148, 176
   group names, single use 125–126
   group reputation and criminality 125,
      131, 182, 185–186
   group response to mimicry 133–134
   group signatures/trademarks 115–116,
      137, 140–146, 183–184, 190–199
   infiltration by groups 229–231, 233–235,
      241
   infiltration of groups 62–63, 66–69, 72,
      118
   intelligence gathering 201–202

Internet and 50–51
passwords 189
as propaganda by the deeds 114–115, 124
recruitment 65–66
signalling 187–202
vetting 76–77
see also assassinations; bombings and
  hoaxes; civil wars; extortion;
  kidnappings
Polyaenus 1–2, 13–16, 19
Potere Operaio 116, 119, 123–124, 140–141
Prima Linea (PL) 116, 125, 129–130,
  132–135, 139 n. 14
Primo Reparto Comunista Combattente del
  Fronte Operazioni Studi Informatica
  Militare 141–142
prison escapes 95–96
Proletari Armati Per il Comunismo (PAC)
  132–133
propaganda, jihadi 76–80

Q
Qaida, al- 21–22, 51–52, 62, 64–66, 77, 79,
  237–238
  AQI (Al Qaeda in Iraq) 236–237, 239

R
Real IRA (rIRA) 91, 103–108
Red Brigade *see Brigate Rosse*
Red Hand Commando (RHC) 97t, 101–102
Red Hand Defenders (RHD) 97t, 101–102,
  106
Reina, Michele 139 n. 14
religious differences 157
Reparti Comunisti Combattenti (RCC)
  141–142
reputation 7, 10, 21–22, 56
reputation systems 74–76, 75f
RHC *see* Red Hand Commando
RHD *see* Red Hand Defenders
Richard I, King of England 16–17
Riggs, Bryan Mark 18, 152–153
rIRA *see* Real IRA
Robin Sage Experiment 55–56
Romas 169–170
Ronde Proletarie 134–135
Rosen, Nir 214, 239
Rossmiller, Shannen 63–64
Rote Armee Fraktion 141–142
Rowan, Brian 107–108

Royal Ulster Constabulary (RUC) 91–93
Rumsfeld, Donald 223–224 n. 15
Rusesabagina, Paul 159–162
ruses de guerre 13, 219–220, 223–224
Russia, bombing of apartment blocks (1999)
  229
Rwanda genocide 45–46, 158–171
  anti-mimicry measures 166–169
  bribery 162
  cultivating Hutu friendships 162
  flight 160–161
  hiding 161–162
  mimicry, absence of 165–174
  passing and mimicry 162–163, 171
Rwandan Patriotic Front
  (RPF) 158–159

S
Sawt al-Jihad 62
Schelling, Thomas 176
Schmitt, Eric 62, 64, 77
Scullion, John Patrick 102–103
Search for International Terrorist Entities
  (SITE) 61
Second World War 18
  Italian resistance 117
  Jews in Germany Army 18
  Jews mimicking non-Jews 10, 45–46
  perfidy 225
  soldiers wearing enemy
    uniforms 219–220
Secret Intelligence Service (SIS) 100
Semwaga, Félix 162
Shahar, Yael 70
Shanker, Thom 62, 64, 77
Shibboleths 19
Shumukh, al-, forum 53f, 54, 59–60, 68–69,
  75, 75f
signalling games 8–12, 18, 42, 47; *see also*
  mimicry
signalling theory 8, 11, 47, 115–116
  costly signs 8, 56–57, 74, 80, 115–116,
    137–139, 176, 187–188, 191t, 190–200
  signalling knowledge 11–12, 80–81,
    198–200
  see also code words
Skorzeny, Otto 219–220
sleeper cells 216–217, 229–230
Smyth, Anne-Marie 98–99
Snowden, Edward 61

social media 81
Somalia, conflict in 158, 166
Sophocles 224
Spence, Michael 29, 46–47
spying 5, 13–14, 19, 22, 50–51, 57, 59, 61–64,
  66–67, 69, 71–72, 99–100, 222–223 n.
  13; *see also* surveillance, digital
Srebrenica massacre 224
Straus, Scott 159 n. 12, 159–160 n. 13
Sun Tzu 12–15, 224
surveillance, digital 51, 57–60, 67; *see also*
  spying
Syria war 81
Sztompka, Piotr 56, 74

T
T'ai Tsu, Emperor 14–15
Taliban 21–22, 68
Taverner, William 12–13
tax returns, falsifying 44
Telegram 52, 79–80
Terranova, Cesare 129–130
terrorism *see* political violence
Terrorism Act (2006) 64
Theognis 20
Timarchus 15
Titterton, John 13, 16–17
Torres Soriano, Manuel 67–68
trademarks 183
Traversa, Martino 131, 134
Troubles, the *see* Northern Ireland conflict
trust, interpersonal 51, 55–58, 63–71
Tsouli, Younis 61
Tumblr 81
Turbay, Diana 22 n. 15
Twitter 52, 79–81

U
Ukrainian war 19
Ulph, Stephen 70
Ulster Defence Association (UDA) 91–92,
  96, 97t, 101–102, 105–106
Ulster Defence Regiment (UDR) 91
Ulster Freedom Fighters (UFF) 91, 99,
  101–102, 105
Ulster Protestant Volunteers (UPV) 90, 105
Ulster Volunteer Force (UVF) 91–92, 96,
  97t, 98–99, 101–103, 105

V
Valpreda, Pietro 118 n. 4
Veblen, Thorsten 169
vetting agencies 74, 76–77
violence, political *see* political violence

W
Wahishi, al- 64–65 n. 20
war
  mimicry and 12–20
  perfidy and 206, 222–229
  see also civil wars; Second World War
Weisburd, Aaron 66–67
Weitzman, Lenore 152–153
Wheeler, Everett L. 13–14
Wiener, Erez 218
women
  political violence and 131, 195
  pretending to be men 17–18
  survival strategies 163–165, 172–173 n. 23

Z
Zamili, Hakim al- 235
Zawahri, al- 64–65 n. 20
Zelin, Aaron 54
Zopyrus 15